IRRIGATION AND WATER RESOURCES ENGINEERING

To my wife—Lorna Renée

IRRIGATION AND WATER RESOURCES ENGINEERING

HENRY OLIVIER, C.M.G.

Pr.Eng., Ph.D., D.Eng., D.Sc.(Hon.), F.I.C.E., F.ASCE.,
M.(SA)I.C.E., F.R.Met.Soc.

EDWARD ARNOLD

Printed in Great Britain by
The Camelot Press Ltd., London and Southampton

Foreword

How does one become an irrigation engineer? What is involved? The answer is that traditionally it calls for long practical training in the design, construction and operation of irrigation works. The training grounds in Indonesia, Iraq, the Indian subcontinent, Northern Africa and the U.S.A. have, in the past, provided many famous 'irrigation engineers' by the process of putting young civil engineers through a postgraduate apprenticeship.

Under modern conditions there is a need to adopt a wider outlook. In the last twenty years an appreciation of the problems generated by the population explosion, and its relationship to the vital agricultural input of water, has indicated on a global scale the need for initiating postgraduate courses devoted to the study of water resources in the broadest sense.

The term 'irrigation engineer' perhaps connotes an unduly narrow range of activities. Irrigation is concerned specifically with agriculture but that in itself embraces other important disciplines such as soil science, animal husbandry and hydrology. Furthermore the impressive growth of industrial and domestic water requirements, the increasing need for flood protection and a fuller appreciation of the national benefits accruing from facilities for recreation spelled the demise of the single purpose project and ushered in the era of multipurpose development. More recently we have come to place emphasis on economic and financial evaluations and on the optimization of project formulation and designs.

Thus water resource engineering has become a wider and more complex subject tending towards concepts of catchment (or integrated) engineering and regional development. It involves research into many factors including more efficient water distribution and use in relation to the land.

It was in the light of such considerations that the University of Southampton initiated specialized postgraduate courses. In view of his wide experience and knowledge of this subject the Author was invited to become an external examiner and to deliver annually a series of general lectures presenting a philosophical approach to principles of water resources engineering.

I am sure that these notes, in published form, will prove helpful to engineers and others in many agencies throughout the world. They should serve to stimulate the interest of young engineers in the challenging issues involved in seeking to promote a better living standard in developing countries.

T. A. L. PATON C.M.G., F.R.S., F.I.C.E.
President
The Institution of Civil Engineers
London SW1

'If you cannot see the Tigris in the drop of water and cannot visualise the Whole in the Part at which you are looking it is not a Discerning Eye but a Child's Gaze that you are bringing to bear on the problems around you.'

Ghalib

Preface

Although the practice of irrigated agriculture is as old as civilization itself, modern irrigation engineering can be said to have begun with the work of the engineers of the British East India Company in the first half of the nineteenth century. The greatest step forward from the traditional to the modern took place between 1836 and 1850 when Sir Proby Cautley embarked on the Ganges canal project while Sir Arthur Cotton was constructing the first permanent control works on the Cauvery and Godavari rivers in south-east India.

From then onwards engineers developed the irrigation systems of the Indian sub-continent and also used the experience acquired there to modernize and extend the traditional irrigation works of Mesopotamia and of the Nile Basin. Between the two world wars an outstanding example of their work was the Gezira project in the Sudan.

Up to the 1950s successive generations of irrigation engineers received their practical training overseas and usually spent their working lives with the technical civil service in different parts of the old Empire. This pattern was abruptly changed with the emergence of a Commonwealth of independent countries, each understandably anxious to staff its civil service with its own nationals. As a result, a number of senior irrigation engineers returned home and lent a great impetus to the activity of consulting firms in this field. But, at the same time, the training of young irrigation engineers became more difficult and a real danger was sensed that the continuity of experience could easily be lost.

The 1950s also saw an increased awareness of the possible consequences of the population explosion on the world's future food supplies. With the population density growing most rapidly in under-developed, semi-arid regions, an increase in the agricultural product was seen largely to depend upon timeous and secure supply of water to the land.

It was against this background that the University of Southampton inaugurated in 1964 facilities for postgraduate studies in Irrigation Engineering. This decision stemmed from the growing awareness of the need for trained personnel to conceive, design, construct and operate water control projects, particularly in the under-developed territories, so as to achieve optimum benefits from available water and land resources.

The course, leading to a Master of Science degree, was designed to suit both those who had been practising engineers for some time and those who had freshly graduated.

Studies are taken to advanced level in the relevant aspects of hydrology, hydraulics, soil mechanics and structural engineering. At the same time the field of study is not limited to engineering subjects, but is intended to develop an overall appreciation of the problems involved in land and water development projects with emphasis on economic evaluation criteria.

Early in 1964, Dr. J. R. Rydzewski, on behalf of Professor P. B. Morice and the University, invited the author to become external examiner for the course for a period of four years and to deliver annually a series of lectures covering the broader aspects of irrigation and land drainage engineering. The objective was to demonstrate generally the wide spectrum of disciplines involved in a practical approach, under modern conditions, to water resources development with special reference to irrigation.

By agreement of the University Authorities the basic notes for the series of nine lectures delivered is reproduced in this volume in the hope that interest may be stimulated in a wider, international, and not necessarily academic, field.

Acknowledgement of special indebtedness is due to the following:

Sir Alexander Gibb and Partners, Consulting Engineers, London, and to Dr. J. R. Rydzewski, Ph.D., F.I.C.E., Department of Civil Engineering, The University, Southampton, Hampshire, for facilities, help and encouragement during the period of launching of the Course and in the preparation of this work.

Mr. E. P. Delany, M.Sc.(Eng.), F.I.C.E., F.A.S.C.E., for his assistance in the final marshalling of the data and checking of calculations and graphs and particularly for permission to quote from his thesis for the degree of Master of Science in Engineering at the University of the Witwatersrand, completed in 1967: *Evaluation Techniques and their Role in Water Resources Development Planning.*

HENRY OLIVIER

Contents

List of Tables

List of Plates

Conversion Table

(According to 'The Use of SI Units' British Standards Institution P.D. 5686, January 1969)

English	Metric
1 inch	$=25\cdot4$ mm (millimetres)
1 foot	$=30\cdot48$ cm (centimetres)
	$=0\cdot3048$ m (metres)
1 mile	$=1609\cdot31$ m
	$=1\cdot609\,31$ km (kilometres)
1 pound (lb)	$=0\cdot4536$ kg (kilogrammes)
1 long ton (2240 lb)	$=1016\cdot05$ kg
	$=1\cdot016\,05$ metric tonnes
1 acre	$=4047$ m^2
	$=0\cdot4047$ ha (hectares)
	$=4\cdot047 \times 10^{-3}$ km^2
1 square mile	$=2\cdot5899$ km^2 (square kilometres)
1 cubic foot	$=0\cdot0283$ m^3 (cubic metres)
1 cubic yard	$=0\cdot7645$ m^3
1 gallon	$=4\cdot543$ l (litres)
1 acre-inch	$=102\cdot6$ m^3
1 acre-ft	$=1232$ m^3
1 MAF $=10^6$ acre-ft	$=1\cdot232 \times 10^9$ m^3
1 Rupee	$=0\cdot21$ U.S. dollars
1 Maund (80 lb)	$=36\cdot3$ Kg
1 cusec $=1$ ft^3/s	$=28\cdot3$ litres/s
	$=0\cdot0283$ m^3/s
1 lb/acre	$=1\cdot120$ kg/ha
	$=0\cdot112$ t/km^2
1 Maund/acre	$=89\cdot5$ kg/ha
1 Rupee/acre	$=0\cdot519$ dollar/ha
1 Rupee/acre-ft	$=0\cdot259$ dollar/1000 m^3

1 Introduction: A review of trends in irrigation engineering

Irrigation engineering is concerned with the conception, planning, construction and operation of projects to provide that indispensable catalyst for agricultural development, water. It is not a single subject, but rather a complex of subjects, overlapping or tangential to each other. It embraces technical (*i.e.* scientific), administrative, economic, political and, above all, intensely human fields. The entire complex of subjects is dynamic in character. The compelling dynamic forces originate from the integrated effects of the population explosion, accelerating industrial expansion, and loss of land from wastage, erosion, and salinity encroachment. These forces have served to focus attention on regional and world water standards. Accordingly, we are now becoming increasingly aware, first of the need to evaluate correctly the complicated hydraulic time disciplines relating requirement and availability at a given moment in time, and secondly of the economic and financial constraints arising from competing water needs and fluctuating budgetary policies.

Worldwide trends of development, both agricultural and industrial, indicate a strong case for the constructive appraisal of irrigation engineering as a factor in promoting balanced development, with special reference to nutritive and economic viability.

BRIEF HISTORY OF IRRIGATION DEVELOPMENT

In earliest times the purely agricultural communities tended to settle in the better-watered regions, *i.e.* in areas of highest or most reliable rainfall and along the main rivers.

There is, however, evidence that irrigation was practised from prehistoric times. Egypt claims to have had the world's oldest dam, built some 5000 years ago to store water for drinking and irrigation. Basin irrigation introduced in the Nile valley around 3000 B.C. still plays an important part in Egyptian agriculture. The ancient skills of the Persians in extracting water from the screes of the foothills by means of horizontal wells is well known.

In an era of slow communications, neither qualitative nor quantitative appreciation of the benefits of irrigation was possible. It is very likely that degeneration of the sophisticated irrigation systems of the Lower Euphrates Valley, of which traces remain to this day, was due as much to administrative neglect and salinity as to ravages following hostile invasions.

At the beginning of the nineteenth century, irrigation patterns in the sub-continent of India and Pakistan had decayed, and British engineers in that area observed and were impressed by the vicious effects of famines resulting from unpredictable and recurring droughts on a population dependent on rain or river flood agriculture.

The Industrial Revolution of Europe ushered in a new era of invention, discovery, mechanization, and improved communications. There followed a more rapid redistribution of population—from rural areas to cities and from country to country—and consequently a sharper realization of the term 'standard of living'. The effects of recurring famines on populations were observed, reported, and interpreted. 'Protective' schemes designed to mitigate the effects of drought were devised and undertaken irrespective of ranking based on criteria of financial viability. Other schemes, termed 'productive', were undertaken according to their financial viability. The agricultural sector acquired a new status; it became the subject of pilot experiments and the new tools of science were applied to the conception and construction of large-scale perennial irrigation schemes in order to stabilize agriculture and eradicate the scourge of famine. There dawned an increasing awareness of the fact that shrinking agricultural communities would need to support increasing non-agricultural communities, thereby incurring expanding obligations over and above those of self-sufficiency. Moreover, the agricultural sector developed its own industrial 'wing' by the production for export of cash crops as distinct from subsistence crops.

These pressures resulted in the initiation of the vast classic irrigation projects of Egypt, Indian, Pakistan and Iraq. It has been estimated that during the nineteenth century the world's irrigated areas increased from 8 million to 40 million hectares (20 million to 100 million acres)[1] mainly following single-purpose projects.

During the first half of this century the pace of irrigation development accelerated sharply. The total world area under irrigation at present is estimated to be of the order of over 160 million crop hectares (400 million crop acres). The trends for the Punjab regions of India and Pakistan and for the 17 Western Reclamation States of the U.S.A. are illustrated in Fig. 1.

Generally the process of evolution in respect of irrigation water as an input in agricultural practice appears to be characterized by three distinct phases:

(a) During the pioneering or low pressure stage the people gravitate naturally to the best water sites and climatic regions. During this phase the firm requirements are met by the 'run of the river'. This phase is particularly important because of the establishment of a system of water rights which, in later phases, generates constraints to development in the best interests of regional and inter-regional needs.
(b) During the intermediate or transition stage the run of the river supplies can no longer meet the growing needs of the population. Storage schemes are resorted to, generally of the single use type. Since pressures are still low and a wide choice is available of good storage sites, little attention is paid to the needs of other users and to the 'value' of water. Thus the principles of the greatest benefit to the greatest number are generally ignored and projects are evaluated on a strictly 'financial', as distinct from an 'economic', basis.
(c) The high pressure or multipurpose stage is reached when demands resulting from population pressures have increased to the point where it is recognized that water resources must be developed according to principles which ensure the best overall use of the scarce commodity. In planning for development attention has now to be paid to

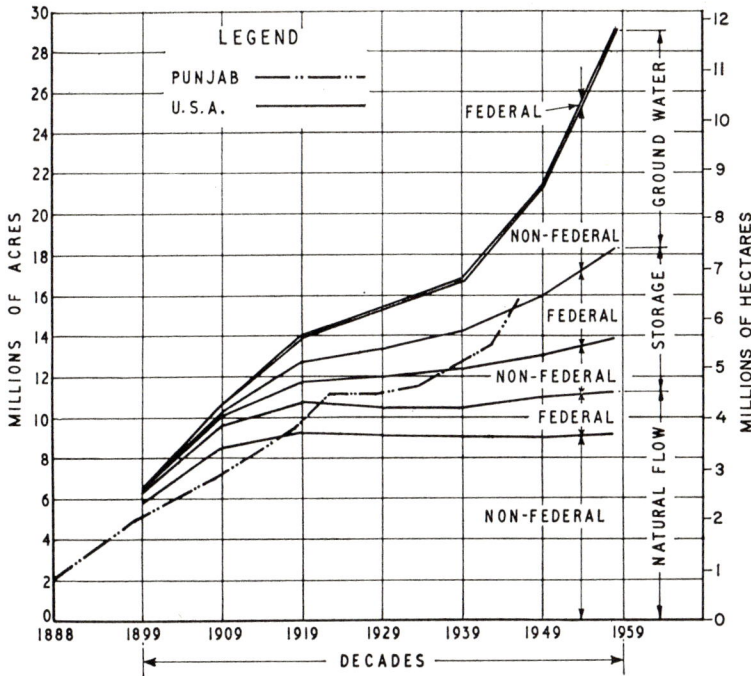

Fig. 1   Comparison of trends of irrigation development for the Punjab and Western Reclamation States of the U.S.A.

competitive pressures and multipurpose uses. There is a sharp increase in emphasis on efficiency of use and/or re-use and on relative production values for various uses of water. At this stage problems are aften accentuated by virtue of accumulated effects from malpractices in the past which make it necessary to face reclamation as well as construction.

Towards the middle of this century the term 'supplementary' appeared increasingly in literature on irrigation. One use of this expression related to the need to supplement natural river flow by storage projects, and subsequently to supplement regulated river flows from underground sources (Fig. 1). The second arose through the increasing appreciation of the effects of time disciplines imposed by climatic and other factors on crop yields, which led to deliberate policies to supplement precipitation in the fields by carefully phased make-up irrigation waters.

Occasionally confusion is caused by the loose use of the term 'supplementary irrigation' to apply to adjacent, *i.e.* additional irrigated lands, as distinct from those changed from dryfarm or rainfall practice.

The irrigation factor in the economic analysis of agricultural production has received particular stimulus from the following developments over the past fifty years.

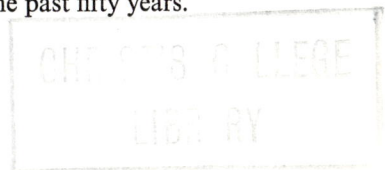

(a) The effect of two world wars is focusing attention on relative standards of living which has resulted in greater collaboration between agriculturalists, engineers, scientists, economists and financiers in all phases from conception to execution of the many projects competing for the relatively scarce financial resources available.

(b) The increased mechanization of farming, extensive use of fertilizers and improved methods of pest control have significantly raised the benefits of output per man hour, and yields per hectare and per unit of water. Advances in the design of earthmoving plant have made it possible to contemplate dam construction even on sites with indifferent geological conditions. Consequently, modern irrigated agriculture is being included in the capital intensive sector of the economy.

(c) The conception and construction of large multipurpose projects with power generation and/or flood control as their main objectives create the storage facilities to provide 'new' water for firm irrigation development. The Snowy Mountains Project of Australia is a good example.

(d) Techniques and 'design tools' have improved as a result of research, analyses, correlation and rapid dissemination of agricultural statistics, and lately the application of computer techniques to systems analyses.

Among immediate byproducts has been the increasing application of worldwide, as distinct from local, experience to the formulation and operation of projects. As a corollary, there has been a tendency for critical examination of irrigation efficiency[2] and for concerted efforts to be made to establish meaningful and effective criteria for the assessment of the economic viability of projects.[3] [4]

IRRIGATION AS A FACTOR IN BOOSTING CROP YIELDS

It is a difficult matter to assess exactly the physical and financial benefits directly attributable to irrigation because of the many other factors that influence this. Indeed, fertilizer is often best applied because of, and by irrigation; similarly an increase in on-farm investment in such items as machinery or storage facilities frequently follows the availability of a guaranteed firm supply of water. It is, however, accepted that in many of the arid regions production is not possible at all without irrigation, and in the humid regions recent statistics on a 'with' and 'without' basis reflect many of the advances made in farming techniques under other headings. The extensive statistical data available and the general conclusions and findings reached by government or private research organizations do, however, provide a reasonably reliable indication of the importance of irrigation as a factor in the boosting of crop yields.

Results from recent trials in Britain[5] for main-crop potatoes have shown that on the light gravel at Gleadthorpe the mean increase following irrigation over five years was 8·3 t/ha (3·3 ton/acre), or 40–45%, and in the West Midland regions the increase was more than 12·5 t/ha (5 ton/acre). Work at Reading has shown that water alone increased the yields of dry matter by 42% and 46% respectively in 1955 and 1956.[6]

Average yield/hectare for irrigated and non-irrigated crops in the seventeen conterminous Western States and Louisiana, taken from the U.S. Census of Agriculture for

1959 and from the U.S. Department of Agricultural Statistics for 1962[7] are summarized in Fig. 2, which also indicates the national averages for the whole country. It will be noted that the yields without irrigation are not markedly different from the national averages.

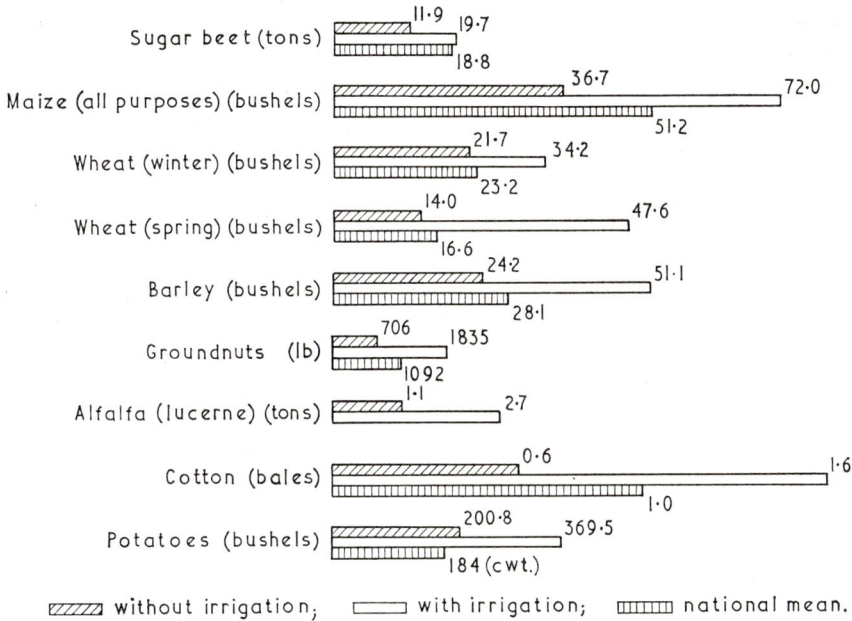

Fig. 2   Mean crop yields in seventeen conterminous Western States of the U.S.A., 1959.

The Kerr Select Committee Report[8] to the U.S. Senate in 1959, based on an exhaustive national survey, provides firm evidence of the beneficial effects of irrigation. Table 1 shows a comparison of irrigated and non-irrigated lands.

Table 1
COMPARISON OF CROPLAND AND PASTURE OUTPUTS FOR IRRIGATED AND NON-IRRIGATED LAND
(also see Table 1(a) in the Appendix)

Location	Value of production per hectare in 1947–49 dollars 1954 statistics			
	Cropland output per hectare		Pasture output per hectare	
	Irrigated	Non-irrigated	Irrigated	Non-irrigated
Eastern U.S.A.	830	158	46·50	9·11
Western U.S.A.	358	79	59·60	2·98
U.S.A.	385	124	59·15	4·41

Note: In most eastern regions of the U.S.A. irrigation is used primarily for high-value crops and the output per unit is thus much higher than for western regions.

The Report concludes that because of the high yields secured on irrigated land, the small percentage of irrigated land that is idle, and the fact that it is not necessary to summer-fallow irrigated lands in the manner practised in extensive sub-humid areas, one acre of irrigated cropland will produce on a national average the equivalent value of three acres of non-irrigated cropland. These figures are, of course, masked to a certain extent by the fact that the use of fertilizer and other factors have been taken into account. Moreover, the statistics in Table 1 also relate to livestock products and fruits and vegetables. Irrigation often results in the switching of land from the production of crops in surplus supply to the production of crops in demand. Generally, the improvement in yield resulting from irrigation alone is somewhat lower.

The broad conclusions to be drawn from accumulating experience may be summarized briefly as follows:

(a) Supplementary irrigation not only increases yield/hectare but also stabilizes the entire basis of farming. In the U.S.A., a conservative estimate shows that on a average national the yield/hectare with good practice is approximately doubled by changing from rainfall basis (dryfarm) to supplementary irrigation. The uncertainties of crop production in semi-arid regions, dependent on rain cultivation or inefficient irrigation practice, have been illustrated by Beauchamp[7] using data from Illinois Agricultural Statistics for yields of maize from the period 1900–57 (Fig. 3).

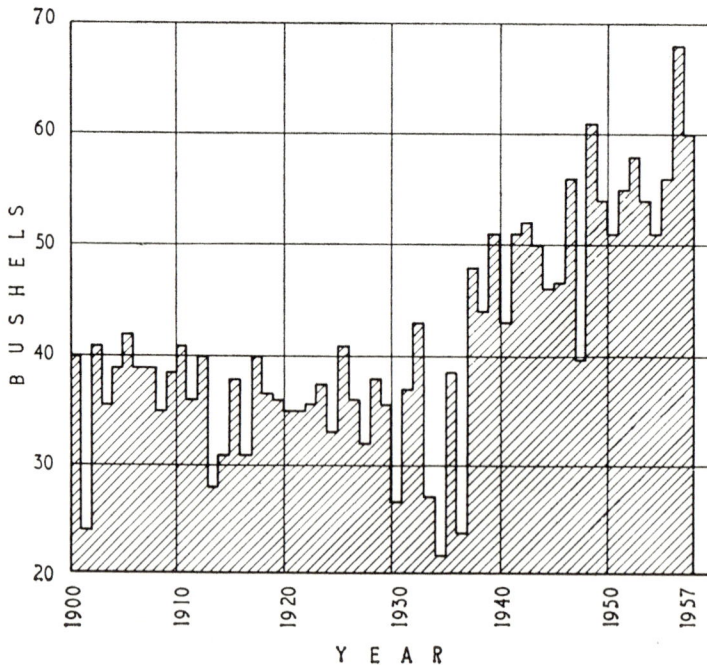

Fig. 3 Mean average maize yield in Illinois.

(b) Overall irrigation efficiency tends to vary greatly throughout the world and is low, of the order of 30% or less. The scale of benefits to be derived from increasing the efficiency of irrigation may be demonstrated by the fact that it is intended to double the existing irrigated areas in the U.S.A. by the year 2000 with hardly any change in requirements for stored and diverted waters, chiefly by raising the irrigation efficiency for the Eastern and Western regions by 10% and 15% respectively.

(c) Irrigated agriculture leads to flexibility and diversity of crop production, which permits shifts in production to meet changing requirements. Thus, in recent years there has been increase in the emphasis on protein and protective foods, which in turn has resulted in a move towards the production of livestock, fruits and vegetables.

(d) Significant increases in the efficiency of irrigation will require, particularly in most of the developing countries, a well-directed and informed campaign of education and demonstration to convince farmers of the practical benefits involved.

NUTRITIVE CONSIDERATIONS

A global statistical review shows a widening gap developing (as regards human needs) between irrigation requirements and supply. Two major factors appear to be responsible for influencing the rate of divergence, *i.e.* growth of the gap. On the one hand, there are the mounting effects of the immense advances achieved in recent times in the field of medical science which have resulted in major reductions in infant mortality coupled with a greater expectation of life in general, particularly in the older age groups, with consequent increases in human consumption in respect of food and water. On the other hand are the limitations placed upon engineering and the allied sciences by financial, political and environmental restrictions.

It is inconceivable that the momentum of medical achievement and of improved living standards should be arrested. Therefore, it follows that there will be no mitigation in the growing challenge to governments in general, and to farmers, agriculturists and engineers in particular, to meet the increasing needs of humanity in the field of agricultural production.

The spectre of the hungry world has introduced new 'currency' considerations into irrigation language; nutritive standards based on calories per head, per hectare, and per unit of water. Accordingly, projects are increasingly tested for nutritive as well as for economic and financial viability.

Optimal combinations of the principal nutrients have not as yet been defined with absolute confidence, and the relationship for a given area between any one or all of these and the units of cultivated land and available irrigation water is a complex problem. Considerable research will therefore be necessary before it is possible to establish reliable links between nutrient yields per hectare in terms of irrigation supplies. Among factors exerting an influence on such a relationship are the different ratios between food plus forage and fibre crops grown in a particular area, the extent of fallow adopted, length of growing season, the extent of double cropping per hectare, the relative importance of calorie, protein and fat nutrients, and the agricultural and irrigation efficiency of the different countries.

The most comprehensive and reliable surveys by country linking nutritive values of food supplies with population and cultivated area are carried out regularly by the Food and Agriculture Organization of the United Nations. These reports draw attention to serious limitations in the preparation of such international balance sheets, one particularly serious defect for the purpose of comparative reviews being that food balance sheets estimate average needs and supplies per head of poplulation and do not reveal the variation of standards within the populations.

However, the calorie standards provided by the F.A.O. are generally accepted by nutritionists as a general, if approximate, standard of measurement. Recent statistics[9] are used in Table 2 to contrast problems and potentials in terms of population and water/land resources for a few selected countries projected to the year 2000.[10]

Table 2
DIET LEVELS IN RELATION TO EXISTING AND PROJECTED IRRIGATED AREAS FOR SELECTED COUNTRIES
(also see Table 2(a) in the Appendix)

Country and projection period		Population $\times 10^6$	National mean daily diet levels kCal/head	Total area irrigated $km^2 \times 10^3$
Egypt	1961	26·6	2530	24·7
	2000	56·0	1800	34·0
Sudan	1960	12·1	2500	8·10
	2000	25·0	2500	18·2
India	1960	442·0	2040	234·2*
	2000	670·0	2300	441·0*
West Pakistan	1960	43·0	1970	68·7†
	2000	65·0	2300	129·4†
U.S.A.	1960	184‡	3100‡	165·9‡
	2000	311‡	3100‡	1132·0‡
Mainland China	1960	640	1900	740·0
	2000	1380	2000	1832·0

Notes: * Includes a considerable proportion of double-cropped square kilometres
    † Includes a considerable proportion of double-cropped square kilometres and relates only to areas irrigated by perennial canals
    ‡ High projection

It seems that in order to ensure adequate food production for minimum diet standards by the year 2000 for China, India and Pakistan an additional acreage amounting to some 80% of the present world irrigated acreage must be brought under irrigation in less than 40 years, a vast undertaking by any standards.

The diet-deficient regions include Latin America, Africa, Western Asia, Communist Asia and the rest of the Far East, in all involving two-thirds of the world population. The nutritional gap in relation to population for these regions is illustrated in Table 3.[11]

This matter is also referred to in Chapter 5.

Table 3
PERCENTAGE OF PROJECTED (1962) TOTAL
NUTRITIONAL GAP BY REGIONS

Region	Population %	Nutritional gap %
Latin America	9	6
Africa	12	6
West Asia	4	3
Far East	42	60
Communist Asia	33	25
	100	100

ECONOMIC CONSIDERATIONS

Irrigation engineers are rapidly learning to distinguish between the economic and financial viabilities of projects. Generally, economic considerations are applied to the assessment of large schemes when the authorities have to decide whether the available resources should be invested in the irrigation project, or whether a greater yield to the economy will accrue from making the investment in some alternative project. The aim of the financial analysis is to show whether the earnings of the project will be sufficient for debt service charges and to pay off initial borrowings, and the extent of profits to be expected.

In large multipurpose projects, affecting vast sections of the national economy, it is the economic analysis which is required with costs and benefits evaluated in terms of real values to the economy. These differ from the values relevant to financial appraisals. For instance, to the extent that there may be distortion in the market prices it will be necessary to resort to 'shadow' pricing in the calculation of economic returns. In cases where farmers pay unrealistic prices for fertilizer because of subsidies, the real cost to the economy, inclusive of subsidy, should be taken. The cost of a project in an economic analysis includes all the items essential to its successful implementation, such as cattle-breeding centres, roads, processing facilities, agricultural buildings, and extension services. In some developing countries supplementary investments relating to the digging of distributaries and field channels on farms are often evaluated as costless in economic terms because the work can be accomplished during periods when the farmer would otherwise be idle.

In many countries the official rates of exchange between local and foreign currencies are only maintained at their official value by means of export restrictions and high tariffs. Thus, the foreign exchange costs of a project may be undervalued in terms of local currency if the official rate of exchange is adopted in the calculations.

The land question is always interesting in large schemes, there being entries on both the cost and the benefit sides. On the one hand there are the costs of land acquisition and resettlement, and benefits forgone from areas flooded by reservoirs. On the other hand, there are the downstream benefits of controlled discharges and levels, and the consequent improvements in land value and progressive easement of flood damage.

With regard to methods for ranking projects for economic viability, the current practice most in favour for comparing costs and benefits is the 'discounted-cash-flow' method. The total investment costs in the project are entered into a cash flow analysis year by year and the net benefits expected from the project are treated in a similar manner. The rate of discount which equalizes the discounted value of benefits and costs at any time is the internal rate of return. More important than the method of calculation is the interpretation of the results. If the internal rate of return is 10%, does this mean that the project is good? With some reservations, providing the internal rate of return is greater than the real cost of money, the project is justified. However, with limited resources it is not possible to undertake all the projects which show a rate of return in excess of the real cost of money. Normally, once the projects are ranked in order, the project with the highest rate of return will be the project to be undertaken. However, other considerations may apply; the project may be the largest of the series and the costs may exceed the capital resources available, or it may be inferior from a financial point of view to another project.

Ideally, a benefit/cost analysis would permit, (a) the ranking of projects in the same field, (b) comparison of projects in different fields, and (c) assessment of the proper expenditure levels for each of a series of programmes. However, in present practice procedures fall considerably short of these objectives, and while perfection in making decisions involving projects that have economic lives of more than half a century is impossible, conceptual inconsistencies in current practice keep the contribution of benefit/cost analysis far short of what it might be. Since it is not possible to assume that projects in different fields with equal benefit/cost ratios have the same economic merit, the technique is still largely restricted to the comparison of projects in the same field. However, a benefit/cost analysis is still the most promising method of evaluation for public expenditure to assist in putting decisions on policy on a much firmer economic basis than is usually possible. McKean[4] has expressed the view that any criterion must be regarded as a partial one only, and one of the most significant partial tests is the maximization of present worth for a given investment budget when the streams are discounted at the marginal internal rate of return. The basic idea is to keep high-value capital from being put to low-value uses.

The very term 'developing country' denotes a dynamic process. No country is ever 'developed' and any particular status is relative, referring generally to the balance or ratio between industrial and agricultural development.

Most developing countries depend on agriculture, which constitutes approximately 60% of their gross national product and provides the basis of living for around 80% of their people.[12] As the country develops (subject to nutritive requirements), and moves up the status graph, the agricultural sector provides the raw materials for industrial growth, and the means for mobilization of capital and the earning of foreign exchange.

At present, for various reasons, irrigation shows a poor comparison with industry in efficiency of water use, as may be seen from Fig. 4 which illustrates from the Kerr

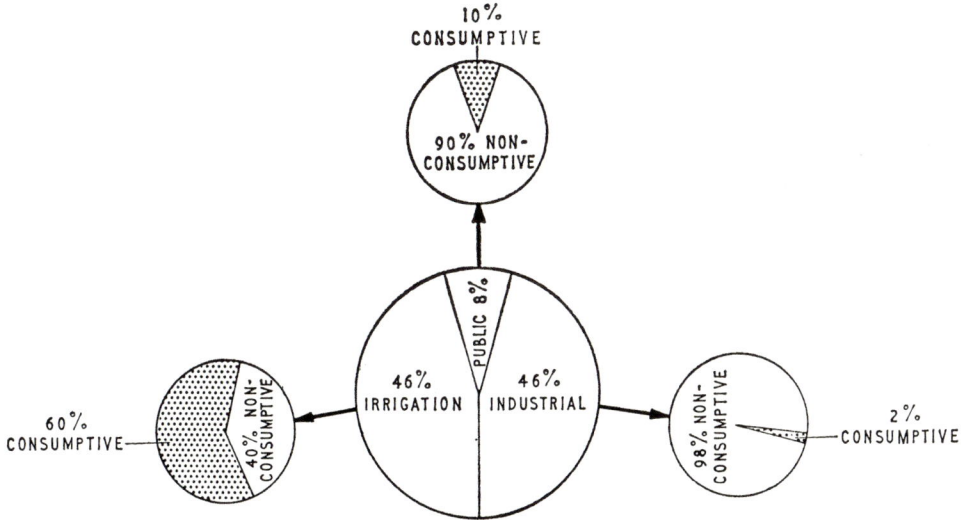

Fig. 4   U.S.A.: National water use.

Committee report[8] the relative national waste factors in the U.S.A. Whereas agriculture and industry each require 46% of the national water use, the consumption of water in agriculture is 30% compared with only 1% for industry.

According to a recent survey for the south-west region of the U.S.A., the relative production values of water have been assessed at from 36·5 to 40·6 per 1000 m³ ($45 to $50 per acre-ft) for irrigation as compared with $3700 to $5000 for industrial purposes.[13] Yet another source gives $17 to $22 for agriculture and $1500 to $4000 for industry.[14] These ratios are undoubtedly higher than the global average but serve to emphasize the problems associated with planning the balanced development of a country's water resources.

Irrespective of whether an analysis is orientated towards nutritive/economic viability or employment potential, the emphasis, whatever the development status achieved, is on increased efficiency in the use of water, particularly in agriculture.

The cost of stored water varies greatly according to site conditions. The general trend is for costs to increase as the more acceptable reservoir and dam sites are taken up, and geologically less favourable conditions have to be exploited. A general indication of costs of storage is provided in Table 4 covering recent developments in Australia, Pakistan, and Central Africa.

Table 4
TYPICAL COSTS OF STORAGE
(also see Table 3(a) in the Appendix)

Country/Region	Project	Storage 10⁹ m³	Cost £ million sterling	Cost per 1000 m³ £
Australia, N.S.W.	Snowy Mountains	2·220*	48·0	21·90
Australia, N.S.W.	Burrinjuck	0·228†	4·0	17·45
Australia, N.S.W.	Burrendong	1·680	14·6	8·68
Australia, Queens.	Coolmunda	0·075	1·9	25·55
Australia, S.A.	Chowilla	5·860	11·2	1·95
Pakistan, Jhelum	Mangla	5·860	165·0‡	28·40
Central Africa, Zambesi	Kariba	44·400**	31·3‡‡	0·71

Notes: * Annual stored/re-regulated waters available and costs attributable to irrigation
  † Raised dam
  ‡ Costs excluding provisions for power
  ** Operating capacity excluding provision for flood storage
  ‡‡ Costs excluding provision for power

IRRIGATION EFFICIENCIES

Applying irrigation water to the fields is like spending taxed income: a metre usefully deployed on the land may mean at least two, and often three, metres stored or diverted, depending on losses in conveyance and application. This is illustrated in Table 5 which gives a summary of conditions in the U.S.A. in 1954.[8] It will be noted that in the Western States, the recovery of losses for downstream re-use amounts to approximately 50% of the diversions.

The approach to improved irrigation efficiencies is governed by two major influencing factors: (a) technical potential, and (b) practical limitations or constraints.

Technical potential

The time disciplines of basic water requirement for any climatic complex and for a given locality can now be determined. Hence there is no longer any point in applying in March the quantities appropriate to July. Except in totally arid countries, irrigation is a make-up process whereby local precipitation is supplemented to meet specific requirements for a particular time with some flexibility in timing provided by the soil moisture storage

Table 5
ANNUAL IRRIGATION WATER REQUIREMENTS PER HECTARE, BY WATER RESOURCE REGIONS, U.S.A., 1954
(also see Table 4(a) in the Appendix)

Water resource region	On farm			Stored/Diverted			Total irrigation requirement m³/ha
	Net required by plant m³/ha	Efficiency of application %	Total required m³/ha	Storage and delivery efficiency %	Total required m³/ha	Estimated recovery of losses %	
Eastern							
New England	616	60	1026	60	1747	20	1541
Delaware and Hudson	822	60	1336	60	2260	20	1952
Chesapeake Bay	1026	60	1746	60	2876	20	2464
South-East	1336	60	2260	60	3800	20	3492
Eastern Great Lakes	822	60	1336	60	2260	20	1952
Western Great Lakes	1026	60	1746	60	2876	20	2464
Ohio	1026	60	1746	60	2876	20	2464
Cumberland	822	60	1335	60	2260	20	1952
Tennessee	925	60	1541	60	2568	20	2260
Upper Mississippi	1232	60	2052	60	3390	20	2980
Lower Mississippi	1336	60	2260	65	3492	20	3082
Lower Missouri	1232	60	2052	65	3182	20	2772
Lower Arkansas, White, Red	1232	60	2052	65	3182	20	2772
Western							
Upper Missouri	1336	45	2980	40	7500	55	4104
Upper Arkansas, White, Red	1643	50	3286	55	5955	55	3594
Western Gulf	1336	50	2672	60	4416	55	2772
Upper Rio Grande-Pecos	2363	40	5860	55	10680	55	6164
Colorado	3184	45	7080	55	12940	55	7600
Great Basin	2158	45	4830	55	8730	55	5136
Pacific North-West	1746	40	4315	60	7190	60	3910
Central Pacific	2670	50	5340	50	10680	55	6263
South Pacific	2876	50	5752	40	11500	55	6780

capacity of the root zone. Advances in the efficient matching of valuable stored water supplies to requirements must come, in the first place, from more accurate assessments of 'effective' rainfall, i.e., that proportion of precipitation which enters and remains within the root zone.

Intensive research is required to determine the influence of natural vegetation cover, litter, and artificial mulches in making a particular rainfall or spray incidence more or less 'effective', having regard to the permeability of the particular soil. Changes in rates of incidence may be considered as analogous to refraction of light through different media (Fig. 5). Also, our knowledge of the practical and limiting conditions under which soil water becomes and remains available to maintain plant life is still rudimentary.

A proper understanding of these processes will lead to improved methods of water accountancy for indents on, or call-up from, stored water. It should also lead to a generally better use of the storage potential within the root zone. Particular refinements foreseen in this direction are in the development of deeper-rooted crops to increase soil storage, and in a water call-up system based on close correlation of allowable moisture stress in the plant rather than on moisture content within the root zone. The difference in approach for a hypothetical crop[15] is illustrated in Fig. 6. It would appear that the moisture stress approach offers better opportunities not only for adjusting water

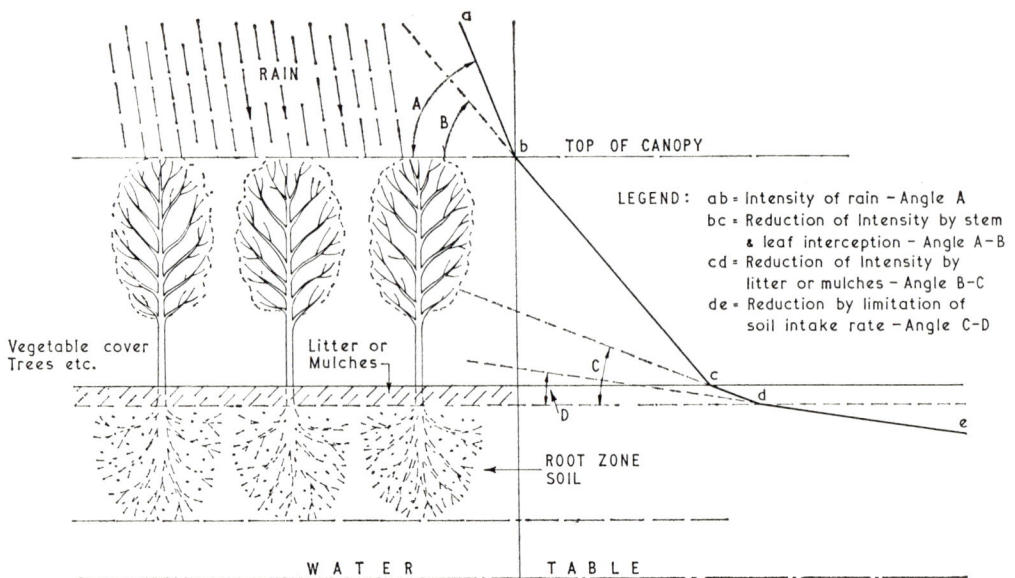

Fig. 5   Rates of availability from precipitation.

indents to take account of cropping factors, but also for more beneficial use of effective rainfall.

Apart from the actual soil intake rates, there is the question of the movement of water tables as a result of excessive total intake, whether from precipitation or irrigation. This has a bearing on drainage requirements and in a positive sense, on the re-use of applied water and the exploitation of underground resources. A diagrammatic illustration of the manner in which irrigation water affects the quality of return water is given in Fig. 7. Return flow studies for the Yakima project[16] indicated that water immediately available for re-use amounted to 20·5% of the diverted water, and that the residual seepage to deep underground storage amounted to a further 20·5%. In territories such as India and Pakistan where perennial irrigation is practised on a vast scale, combined losses of the order of 40% from deep percolation and regeneration seepage constitute major factors not merely as regards the relatively short-term economics of water/land use, but in the progressive qualitative change of water and soils. Preliminary estimates put the annual recharge of groundwater in the northern zone of West Pakistan at approximately $25 \times 10^9 \, \text{m}^3$ to $47 \times 10^9 \, \text{m}^3$ (20–38 million acre-feet) and in the southern zone it is estimated to be about half this amount. It is possible that infra-red photography may, in the near future, provide a means of assessing underground water movements with particular reference to regeneration into streams and re-use of irrigation waters.

There are encouraging indications that significant improvements in irrigation and agricultural practices are possible by deliberate control of local climatic factors by the use of shelter belts. Research work carried out in Russia and the U.S.A. has shown that to

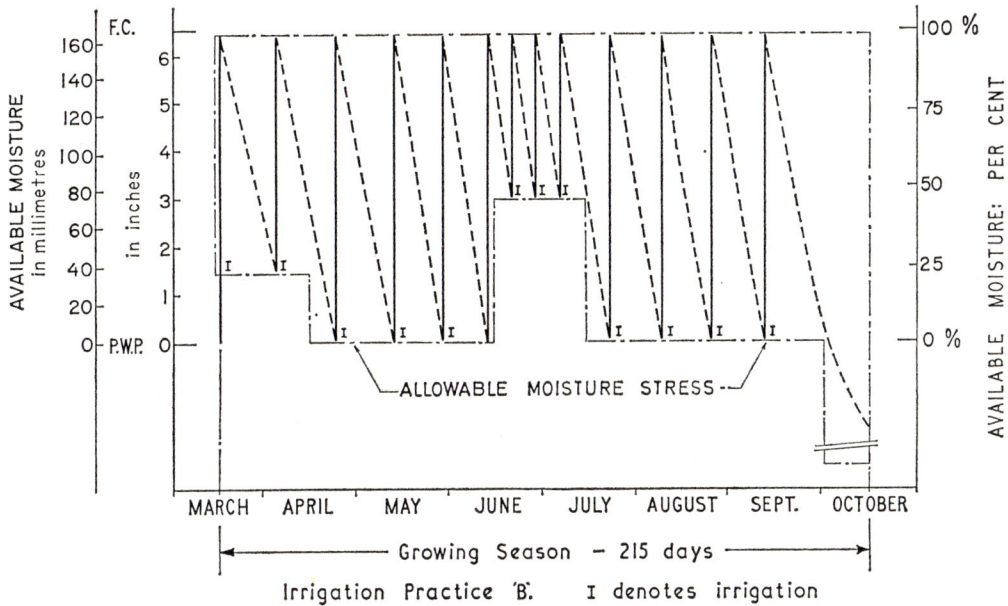

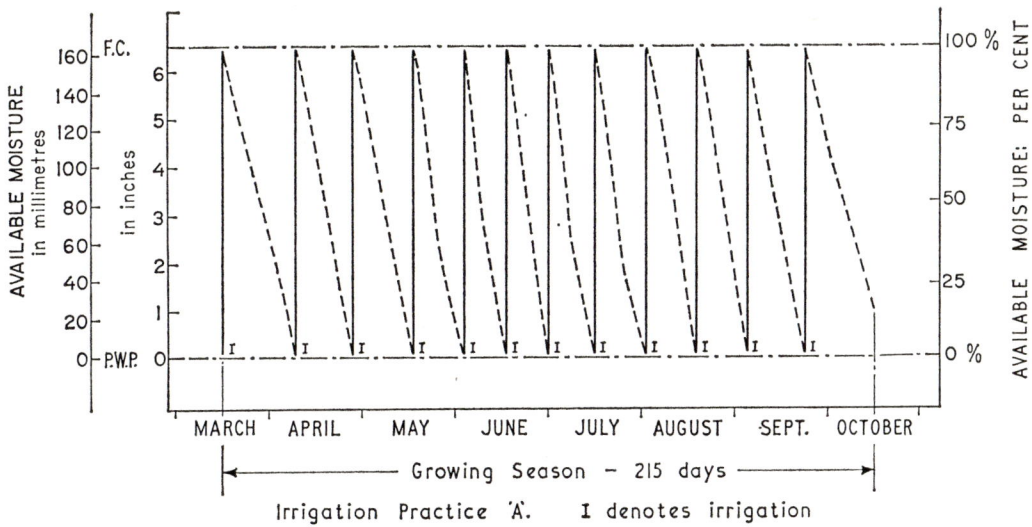

Fig. 6 Comparison of irrigation systems.

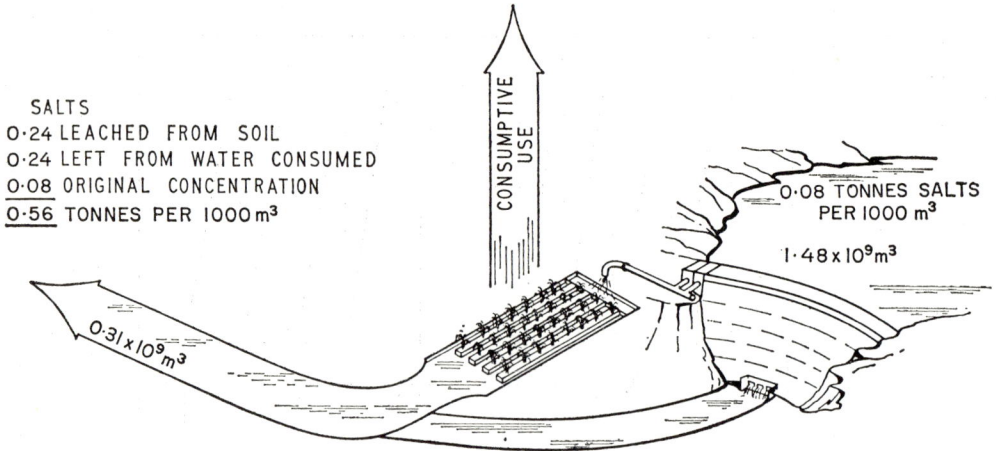

Fig. 7 Typical pollution by irrigation.

leeward of windbreaks, relative humidities tend to be higher and temperatures lower, and yields of various crops are markedly increased when compared with those in adjacent open fields. A general diagrammatic representation of crop yield in relation to leeward distance from a windbreak is shown in Fig. 8.[17]

Calculations based on climatic conditions in Giza (Egypt) show that for an average increased relative humidity of 8% and a reduction of average temperatures by 2% the corresponding saving in potential water requirements is of the order of 20%.

Having regard to aids now available or likely to become available, it should be possible to improve global irrigation efficiency within a few decades by at least 10 to 15%. Indeed, it would seem that technical capacity has outstripped the capacity for implementation because of practical limitations and a chronic lack of basic data.

Practical limitations

At present, less than 10% of the world's area is reliably mapped as regards soil surveys for agricultural purposes. This lack is particularly felt in the diet-deficient regions of the Far East, and in areas of relatively unexplored potential such as Africa and South America.

World-wide meteorological observations with particular reference to rainfall, as conducted at present, make it virtually impossible to assess the effectiveness of precipitation, and hence to evaluate criteria for the improvement of supplementary irrigation efficiency. It is essential to have more automatic recorders. This is not a question of research; it requires merely a change in measuring procedures which could be achieved on a global basis in less than five years.

Technological advances by themselves are quite ineffective if no machinery exists for their dissemination to the masses of farmers and farm workers, in simple practical terms, by demonstrating the economic and other advantages accruing from improved

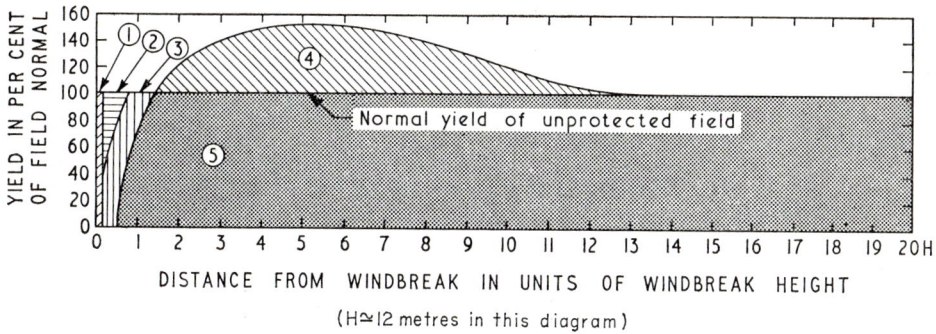

Fig. 8   Cross-section of crop yield on a field leeward of a windbreak.

practices. There is a great mission here for 'non-commissioned officers' in irrigation engineering to assist in the process of education at grass-roots level.

Current economic criteria tend to become vague generalizations when indirect benefits from irrigation projects come to be considered. The evaluation, in an economic sense, becomes as difficult as the assumptions to be made by an engineer when assessing foundation stresses at various depths under a major dam. Moreover, no finite meaningful progress will be made until reliable correlations of statistical data relating nutrition yield levels to units of land and water by region and soil type become available.

One of the greatest practical constraints on irrigation development is the ineffectiveness of current economic criteria with respect to the acknowledgement of residual benefits arising from the durability of major civil engineering structures. The philosophy that works will be completely written off in arbitrary periods of, say, 50 years, based to some extent on the physical life of mechanical plant, make it progressively more difficult to reap the benefits of increased irrigation efficiency, and to plan the integrated use of water resources in multipurpose developments.

Except in the case of relatively minor private-sector works, the trend is for all development of water resources to be multipurpose in character. It is, therefore, necessary to consider relative development periods and possible ultimate use. Space does not permit

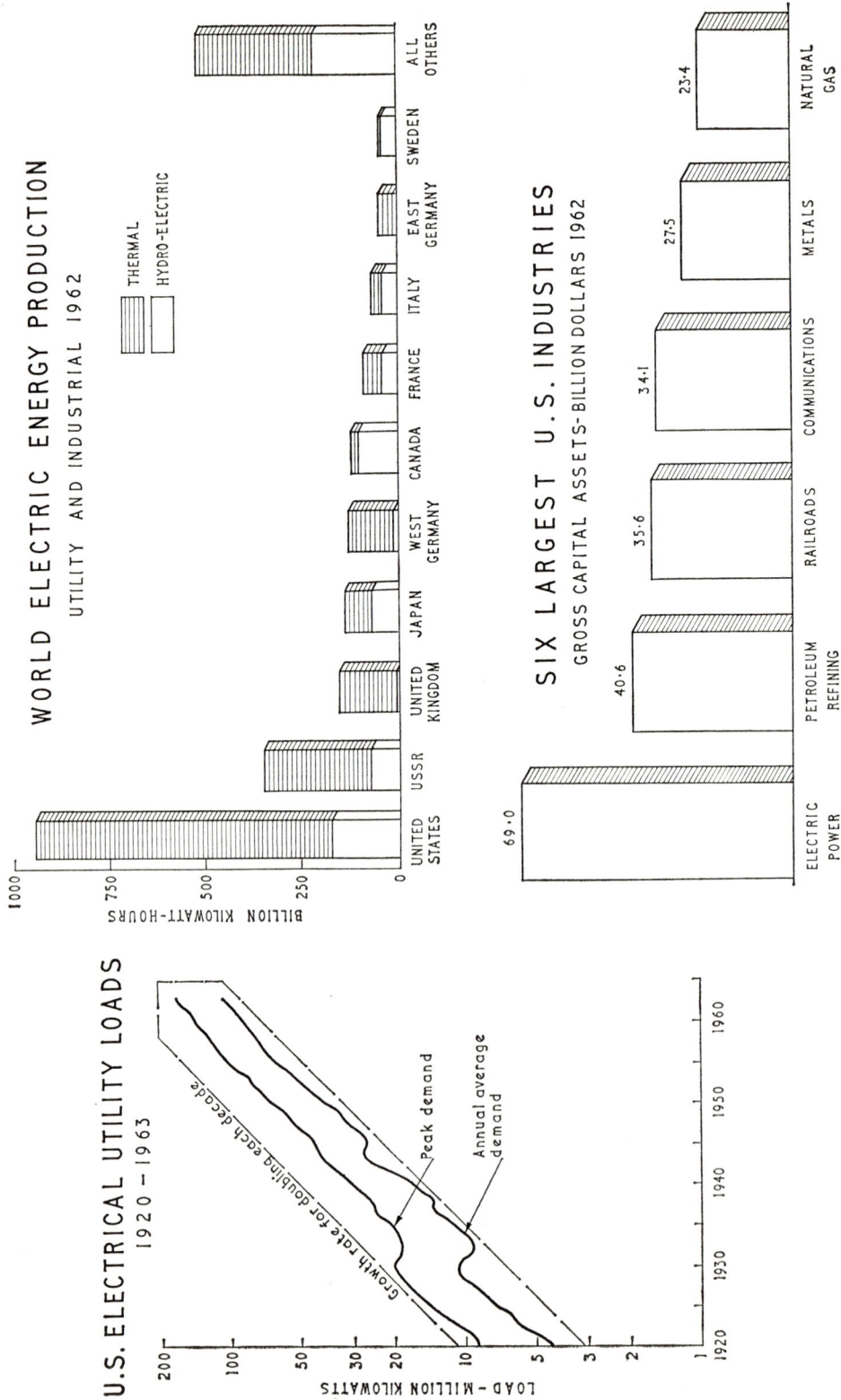

WORLD ELECTRIC ENERGY PRODUCTION
UTILITY AND INDUSTRIAL 1962

THERMAL
HYDRO-ELECTRIC

BILLION KILOWATT-HOURS

UNITED STATES USSR UNITED KINGDOM JAPAN WEST GERMANY CANADA FRANCE ITALY EAST GERMANY SWEDEN ALL OTHERS

SIX LARGEST U.S. INDUSTRIES
GROSS CAPITAL ASSETS-BILLION DOLLARS 1962

ELECTRIC POWER 69·0 PETROLEUM REFINING 40·6 RAILROADS 35·6 COMMUNICATIONS 34·1 METALS 27·5 NATURAL GAS 23·4

U.S. ELECTRICAL UTILITY LOADS
1920 – 1963

LOAD - MILLION KILOWATTS

Growth rate for doubling each decade

Peak demand

Annual average demand

1920 1930 1940 1950 1960

Fig. 9 World electric energy production (1962) and ranking of electric power relative to the six largest U.S. industries.

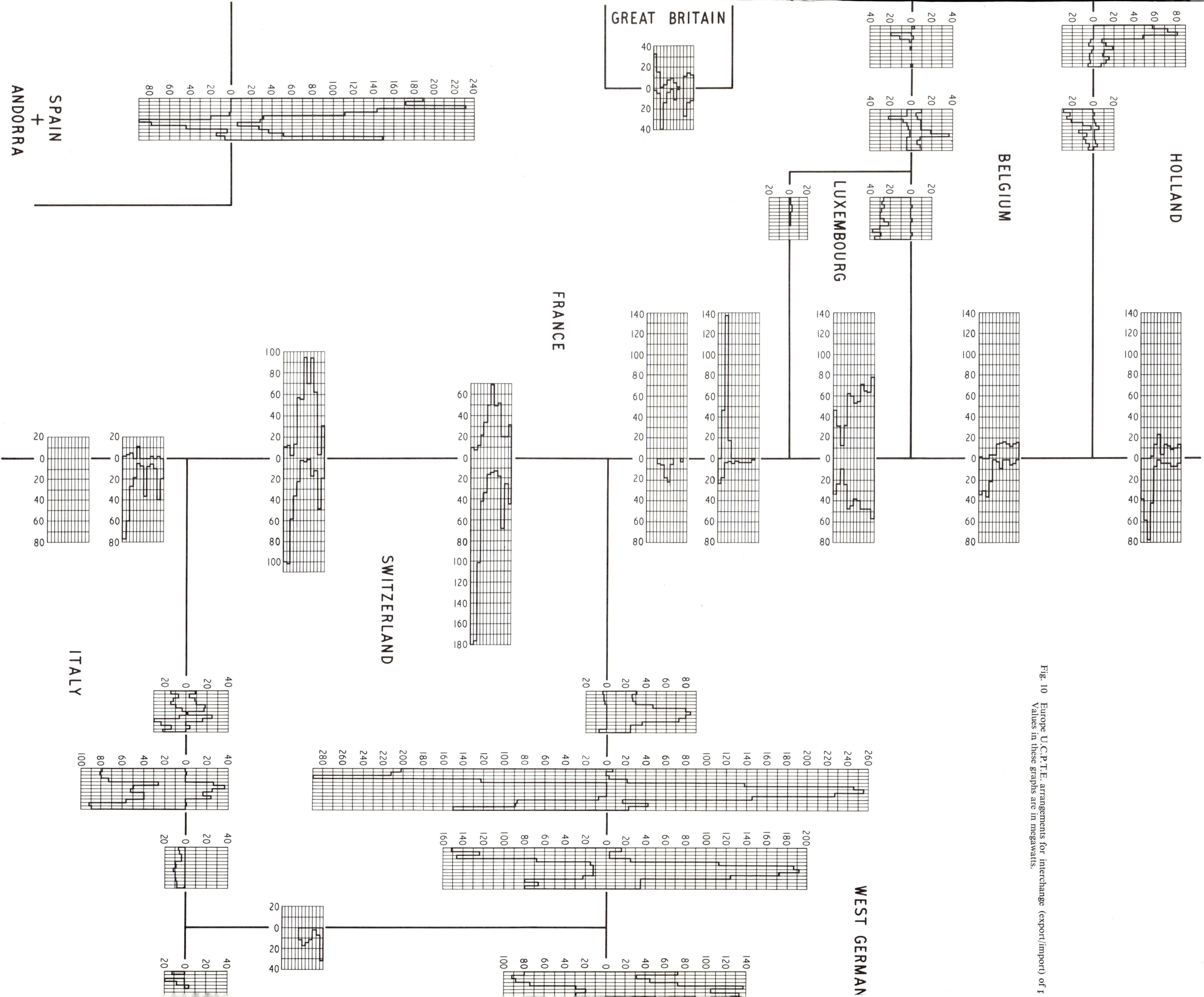

Fig. 10 Europe U.C.P.T.E. arrangements for interchange (export/import) of power. Values in these graphs are in megawatts.

SPAIN + ANDORRA

GREAT BRITAIN

BELGIUM

HOLLAND

LUXEMBOURG

FRANCE

SWITZERLAND

ITALY

WEST GERMANY

more than a brief illustration by reference to the most common partnership in multi-purpose use: irrigation and power generation.

MULTIPURPOSE CONSIDERATIONS: TRENDS IN ELECTRICITY GENERATION

The growing importance of electricity as an 'industry' is demonstrated by the 1963 survey of national resources in the U.S.A., which ranks electricity (with an indicated steady growth rate of 10% per annum for both energy and peak demands) as by far the largest of the six major national industries[18] (see Fig. 9).

The stage is gradually reached in developing countries where transmission networks are adequate to absorb all the energy produced under the load curve, which makes possible the most economic integration of all forms of energy, graded according to availability and efficiency. This has two major economic advantages: export/import relationships between regions become possible whereby surplus power can be supplied for short or long periods to power-hungry neighbouring areas, and it becomes possible to deal effectively with peak loads which, in isolated developments, have to be taken care of by costly and often inefficient peaking capacity. Maximum peak loads arise from weather extremes, industrial activity, and the customs of the population served. A classical modern example of large interconnection is the energy export/import relationship between countries belonging to the Union for the Co-ordination of the Production and Transport of Electric Power of Western Europe[19] (Fig. 10).

Ideally, peaking capacity should have a low initial cost, should pick up load rapidly, operate satisfactorily at partial loads, and be economic in manpower requirements or be capable of remote operation.

A recent trend in the pursuit of greater overall efficiency is for most of the new steam-electric generation plants to be large base load units of 500 MW or more. These are not designed for peaking service since optimum efficiency is only achieved when operated at or near design ratings. The same holds for nuclear sets now coming into service and which, it is forecast, will represent about 8% of total U.S.A. capacity by 1980.

Thus, there is rapidly increasing pressure first to reassess conventional hydro-electric potential which is well-suited to dealing with peaking problems, and secondly to develop pumped storage schemes. The annual cost of installing additional hydro-electric units is generally less than the cost of additional capacity from alternative sources. Moreover, the ability to start quickly and to change power output rapidly makes hydro-electric plant particularly suitable for carrying peak loads and for assisting in the supply of a spinning reserve. Whereas the use of hydro-electric water is non-consumptive, high discharges of water over short periods of time need special studies to ensure against down-stream damage and to produce harmony in overall basin requirements. Recently, hydro-electric plants have been planned and operated for annual plant factors as low as 5%. This trend has already resulted in project sites which were previously considered not feasible for development being found economical as the need for peaking capacity

B

developed. Within the last decade, research in Europe has produced designs for turbines which offer possibilities for economic development of the power potential of sites with heads as low as 3 m to 4·5 m. This holds much interest for the exploitation of river and canal drops, which are common for instance in the vast network of irrigation canals of the Indo-Pakistan sub-continent. Pumped storage schemes are being designed for increasingly rapid rates of load pick-up. The Ffestinniog scheme in Britain has been fully loaded in a period of 51 seconds, and even more rapid rates of loading are contemplated for future installations. These overall trends have created opportunities for conversion of conventional hydro-electric stations to pumped storage schemes when their reservoir storage capacities have been reduced or exhausted by silting.

A review of trends indicates that the paradoxical situation has been reached in which the maximum economic benefit from storage projects is reaped towards the end—and even after—the arbitrarily chosen life of the structures. The repercussions on irrigation and other multipurpose concepts are obvious, particularly when periods of relative development are taken into consideration.

The criteria of economic viability for power and other industrial projects are generally based on post-construction development periods of the order of five years and rarely exceeding ten years. In the case of irrigation projects, the development periods for the attainment of full potential after construction will rarely be less than ten years, and may be much longer.

In regions where population pressures are not yet significant and where the agricultural and industrial rates of development are reasonably balanced, the development period for new irrigation projects will generally be longer than for power projects. In the case of the Vaalhartz irrigation scheme of South Africa, it took nearly 30 years for the 32 400 hectares of the new project to approach their full potential. In India the Rajasthan scheme is due for completion around 1976 and involves a main canal some 692 kilometres long, to command nearly 2 million hectares (approximately 4½ million acres). The development period has been given as ten years. In Pakistan the Khairpur scheme has been planned on the basis of an assumed development period of ten years.

The Snowy Mountains Scheme of Australia was designed to provide hydro-electric power to meet the growing peak loads of the New South Wales and Victoria power systems and, from 1959 onwards, to begin providing regulated irrigation supplies in the Murray and Murrumbridgee rivers up to the total potential of $2·22 \times 10^9$ m³ (1·8 million acre-feet) per annum.[20] The development period is expected to be fairly short since the additional regulated supplies will be used mainly to provide for more intensive development of areas already served by distributaries and the remainder will be available for development of adjacent high-quality land.

Depletion of reservoir storage or diversion potential, whether due to silting or to other causes, has a serious impact on any irrigation project. As it is not usually possible or desirable to reduce farming potential once established, the cost of providing progressive make-up water to maintain supplies must be allowed for in the cost streams of the project analysis.

The concept of changing the priority or emphasis between uses with time is generally becoming increasingly attractive. Thus, in a temporarily power-hungry region, suitably equipped with transmission potential but relatively virgin as regards irrigated agriculture, it is possible to operate the storage project mainly as a power project in the first 20 to 30 years on a financially viable basis. In the meantime, the irrigation potential in the region and also recreational facilities and, with increasing basin development, flood protection can be developed. As the power emphasis wedges-out following technological advances, whether from different sources, such as nuclear plant or from policies of decentralization, the irrigation potential wedges-in. Hence, although irrigation may be regarded as having ultimate priority in terms of projected economic assessment, such priority need not inhibit initial best use of the natural resources.

It is visualized that irrigation efficiency and practice will improve considerably over the next two decades, which will result in fewer requirements at source per unit of land and, to the extent that this is achieved, the flexibility of alternative use and re-use is considerably enhanced.

OUTLOOK

Parallel to a drive towards improved irrigation efficiency, engineers will have to review techniques of dam design. In general the costs of storage are too high on the basis of present design concepts. Greater use will have to be made of *in situ* materials on remote dam sites with hostile geology. In basins where one or two expensive dams cannot be avoided, greater efforts should be made to protect these structures from excessive floods and from loss of service due to silting. This could be accomplished by catchment control, including the construction of a number of cheaper protection dams upstream to regulate the river and to act as silt traps. The use of new techniques and materials should not be ignored, but it will, of course, not be acceptable to reduce the safety criteria of the main structures. This may mean long periods in which prototypes are tested to the satisfaction of all authorities concerned.

Once soil mapping of the world is adequate and meteorological observations made meaningful in an agricultural sense, it should be possible to accelerate the exploration of the possibilities and economics of the development of agricultural zones for world food production, with particular reference to supplementary irrigation. If the same or better yields can be obtained in a humid zone from an application of 1 m of water than from applications of 3 m–5 m in an arid zone, the implication becomes obvious. Where a choice is possible, the trend will increasingly be to locate industries, which have relatively low consumption of water, in the arid regions, and to concentrate on irrigation in the humid regions.

Reviews of resources indicate that some regions can never be self-sufficient as regards food supplies, and others can become so or maintain their nutritive standards only by strenuous efforts to improve efficiency. Thus planning will increasingly be on a regionally co-operative basis: whether by sharing of waters or by export/import (common market) arrangements.

Major irrigation projects cannot, however, be planned on a short-term basis. Economic criteria will need to be refined to take into account forward planning at least 100 years ahead, if projects are to be selected on rational theories of viability, and not on simple 'flash' judgements. This also offers the only prospect of making the best ultimate use of durable engineering structures and provides a long-range basis for planning the development of electrical potential and making the best use of technical advances in hydro, thermal, and foreseeable nuclear generation, and the integration of all resources with expanding transmission systems.

Thus irrigation may well become the corner-stone of multipurpose planning on the basis of long-range economic assessments, and not the balancing factor, or by-product of, other uses which offer short-term attractions on the basis of financial criteria.

2 Irrigation and climate

In many under-developed or new territories, and in some of the older countries, one of the causes of impeded development is lack or misuse of water and, in recent years, much has been written about the necessity for overall planning in the development of water resources. Attention has been focused, mainly by economists, on the necessity for adequate methods of assessing the relative gains to the economy of a country arising from the various possible alternative uses of the available water, and on the real costs associated with its exploitation. This trend towards refinement of the assessment of costs and benefits is being carried on regardless of the very gross assumptions that are often made regarding the actual quantity of water required to produce a unit of benefit.

Recent surveys have brought appreciation of the tremendous advances achieved over the past two decades in boosting agricultural output per unit of land mainly through scientific improvements in fertilizers and seeds, mechanized farming and cultural practices. It has also revealed that there still exists a great gap (estimated by some at 50%) between the level of production per unit possible through full application of science and actual current levels of production on the farms. This would indicate that efficiencies are still capable of improvement as regards output *per unit of water*.

Owing to the widespread nature of operations and the complex ramifications of irrigation practice the economic, as distinct from the financial, advantages of maximizing output per unit of water have not yet become apparent at all levels. The practice of irrigation is still steeped in local tradition. The economic effects of deviation from the disciplines are not so immediately apparent as in the impact of loss of revenue due to failure to meet load duration curves in hydro-electric engineering for instance.

Considerations of improving overall efficiency in irrigation involve understanding of the three distinct sets of hydraulic laws governing the three phases of the operation:
—Storage/conveyance: reservoirs, canals, distributaries.
—Application on the fields into soil storage.
—Consumption from the soil storage reservoir.
Advances in any one phase, such as storage/conveyance, are, or could be, neutralized by mal-operations or continuing waste in the other phases of the irrigation system.

Advances in design and construction methods over the past two decades have resulted in water being stored and conveyed to the point of application with reasonably high efficiency. However, all such gains in efficiency can be, and often are, neutralized by mal-operations in the fields. *It is not efficient to convey efficiently the wrong quantity of water at a particular time, or throughout the season.* Since the key time disciplines are centred at the fields and all refinements relating to overall efficiency must be predicated on the assumption that optimum levels of efficiency are maintained at the fields, it follows that *properly integrated water indents must originate from a technical field command and be transmitted upstream to the storage or diversion point.* Bulk releases from storage according to some arbitrary or traditional pattern cannot maintain, let alone promote, efficient use of resources, and usually result in scrambling or bribing for water in the fields.

The first essential for the preparation of indent components for a given region or crop complex is the preparation, at suitable time intervals, depending on soil storage potential, crop stage, and effective rainfall, of actual field rates of requirement. In this connection, there is need for a simple but reliable method of estimating crop water requirements.

Dalton first enunciated the basic law of nature governing evaporation. Since then there have been many attempts to establish relationships which could be of practical use in estimating both evaporation and/or the use of water by crops, based on different assumptions and parameters. The extent of short-fall in this direction is echoed by the summary of ideal requirements for such a method, given by Soane[21] in Salisbury, Rhodesia:

'(1) Is it capable of application on a world-wide scale, or will it be subject to errors due to local climate variations and hence require local calibration?
(2) Is it designed for the practical use by farmers, or is its use restricted to research workers?
(3) Is it designed to estimate only seasonal or yearly transpiration, or can it give satisfactory results on a day-to-day or week-to-week basis?'

It has not proved easy to meet all the above criteria. Formulae devised under laboratory conditions usually require extensive calibrations when in universal use over broader fields under different climatic conditions. In the case of very complicated formulae the calculations and instrumentation necessary for reliable evaluations are rather extensive, and this is inclined to inhibit use *by engineers* in many of the under-developed territories where instrumentation is not sophisticated and, more important, precludes practical application by the farmer.

In 1953 an empirical approach was suggested by Olivier[22] [23] based on statistical research along lines of thought derived from work done, and correlation factors achieved by, Keeling in Egypt[24] and Briggs and Shantz in the U.S.A.[25] The subsequent results obtained from practical use of the relationship for engineering assessments of crop water requirements have consistently proved to be reasonably reliable when checked against actual supplies on a wide geographical basis in five continents and under different climatic conditions.

The formula is relatively simple and its use involves field readings by relatively unskilled observers of only two or three very simple meteorological instruments. Thus use is made of observations as currently made all over the world and the method is not dependent on the prior introduction of sophisticated scientific instruments or on gross assumptions based on observations in adjacent valleys or regions. Moreover, it can be used to assess daily, weekly, monthly or seasonal water losses through evapo-transpiration, and facilitates the rapid integration of water requirements for a complex irrigation unit and the regular compilation of reliable bulk call-up (indents) from storage consistent with time disciplines imposed by controlling factors such as climate, soil, storage and crop stages.

Having decided upon production objectives, cropping patterns and capabilities of the agricultural community to calculate and manage the farm, it is possible to build up patterns of water demand for comparisons of water availability. From this it becomes possible to derive preliminary ideas of the structures required to store and to convey the supplies to the fields, and to estimate the correct drainage pattern, vertically and horizontally, to avoid raising water tables and the effects of waterlogging. This involves intensive trial and error procedures, and comparison of numerous alternatives. The computer has the great advantage that, having started at farm level and worked back to the diversion site, there is no need to remain wedded to the first concept because the computer can substitute various assumptions in an attempt to optimize the return on capital invested and the phasing of financing needed.

THE METHOD

The rate of basic water requirements in a given locality and under given climatic conditions is determined by the formula:

$$Cu\varphi = cW\varphi \tag{1}$$

where:

$Cu\varphi$ = the basic water requirement in mm/day of a crop-soil unit at the particular locality at latitude φ for the month of study,

c = average depression of the wet-bulb thermometer in °C; exposed in a standard Stevenson Screen for the particular month, at the particular locality,

$W\varphi$ = a dimensionless cyclic (radiation/latitude) factor for the particular month at latitude φ of the particular locality.

Use of the relationship therefore involves only the simplest of meteorological instruments: the wet and dry bulb thermometers. This instrument is in common use in meteorological stations of low order in many parts of the world, due to its simplicity and robustness. It can certainly be used with ease by those responsible for the design or operation of an irrigation system. In particular instances, when the design of an irrigation system is being embarked upon, it may be found that records of the readings of the wet- and dry-bulb thermometers are not available. In this case, the required values of c may be obtained from values of relative humidity and temperature through the use of standard hygrometric tables.

It should be borne in mind that the relationship is based on *averages* of meteorological observations. Unfortunately, a review of world practice reveals that the ideal requirements of true means is seldom easily available. Few stations exist, particularly in areas likely to be explored by engineers, which are equipped with automatic recording apparatus from which true daily or monthly means may be derived by integration of hourly means. It is seldom that more than two readings are taken daily at outlying stations. It has, therefore, been necessary to carry out statistical surveys over a wide geographical field to determine whether reliable approximations to mean values are possible with so few daily readings.

It has been found[22] [23] that a good practical approach to true means of temperature and relative humidity and, therefore, of wet-bulb depressions will generally be obtained with observations made between 0830 and 0930 hours local (or sun) time.

Where calculations for wet-bulb depression c have to be made from published temperatures and humidities the best results under existing conditions would be obtained by a combination of one of the following methods:

(i) Mean temperature = mean of $\frac{1}{2}$ (max. + min.) of daily readings. Mean humidity from mean of readings taken between 0830 and 0930 hours.

(ii) Mean temperatures and humidities both calculated from means of daily readings taken between 0830 and 0930 hours.

The cyclic radiation/latitude factor, $W\varphi$, is obtained from tables. While the form of the above equation is similar to some other formulae in common use: the underlying concept is, however, broader and, it is felt, more logical.

Conceptual consideration of the method

The method is based on two concepts which may be postulated as follows: 'The rate of loss of water from a crop-soil unit to the atmosphere will be controlled by the potential capacity of the atmosphere to absorb the water vapour; and, the extent to which the crop-soil unit will respond to reduce this state of inequilibrium will vary with the level of radiation.'

The first concept in the above postulation relates evapo-transpiration to a meteorological factor which is purposely stated in the passive sense to mark its contrast with the second concept: the degree of response which is held to be dependent on energy availability.

The fact that evapo-transpiration is dependent on one or more meteorological factors has been accepted by most research workers. The process of evapo-transpiration implies a loss of water by a crop-soil unit in the form of vapour. For a crop-soil unit to sustain a net loss of water to the atmosphere, it is necessary for the atmosphere to be capable of absorbing the vapour. Also, the potential rate of loss of water, both by evaporation from the soil and through transpiration by the crop will be dependent on the diffusion pressure deficit between the water in the soil reservoir and the vapour pressure in the atmosphere. It is, therefore, concluded that the concept correlating evapo-transpiration with the absorptive capacity of the atmosphere (as measured by the depression of the wet-bulb thermometer) is the most logical approach to a rational formula.

Accurate measurement of the above capacity and potential in terms of absolute units, if at all feasible, would involve a whole array of instruments. The depression of the wet-bulb thermometer c does, however, provide a measure of the pertinent characteristics of the atmosphere even though its correlation with relative humidity or saturation vapour pressure deficit in non-lineal.

In addition to the above rational considerations, it is postulated that c is a more critical measure than certain other meteorological factors whose measurement is simple and which are commonly used in evapo-transpiration formulae, *i.e.* it represents the integrated total effect of other meteorological component factors.

For instance, a relationship correlating temperature and consumptive use will ignore the depressing effect of an excessively humid atmosphere on the rate of return to the atmosphere of water vapour by a crop-soil unit. A 'temperature' formula would give the same water requirement in an average temperature of, say, 18°C (about 64°F) whether the relative humidity was 40% or 80%, yet it is evident that all other factors being equal, the rate of evapo-transpiration from a crop-soil unit under the first set of conditions will be greater than under the second. Similarly, a relationship, which correlates evapo-transpiration with relative humidity only, would not take account of the effect of the average ambient temperature. A 'relative humidity' relationship would give the same water requirement for a crop growing in an atmosphere at 60% RH, whether the average temperature was 28°C or 17°C. The value of c would, however, be 5·5°C in the first case and 4·0°C in the second case.

With regard to the second concept, it is clearly necessary for energy to be added to the crop-soil-atmosphere system to enable the process of evapo-transpiration to be continuous, otherwise energy required to provide the latent heat of vaporization and to overcome the resistance of the crop-soil system would have to be taken from within the system.

The chief source of energy being radiation from the sun, the available amount varies with latitude and, for a given latitude, with time. Generally, radiation decreases from the poles towards the equator and for a given locality varies on a day-to-day basis depending on the number of daylight hours (or hour angle) and on the declination of the sun.

The energy variations with latitude and time are functions of the physical characteristics of the earth and its relationship to the sun, which vary only slightly from year to year. These may, for practical purposes in this context, be considered as not being subject to change from year to year, and are represented in the formula by the expression $W\varphi$, which is based on the monthly variation of the daily average vertical incident solar radiation under 'clear-sky' conditions. The values of $W\varphi$ tabulated in the Appendix to this chapter are derived from the following equations:

$$W\varphi = Lo/L^2 \qquad (2)$$

where:

$\quad L =$ the value for any particular month of the ratio R/Rv at latitude φ,

and

$\quad Lo =$ the average annual value of the ratio R/Rv at latitude φ.

The ratio R/Rv being obtained from the equation:

$$Rv = R \sin h \qquad (3)$$

in which

$$\sin h = \sin \varphi \cdot \sin d + \cos \varphi \cdot \cos d \cdot \cos t \qquad (4)$$

where:
  φ = latitude of the place of observation,
  d = delineation of the sun,
  t = the hour angle.

The clear distinction between the two factors c and $W\varphi$ is fundamental to the understanding of the suggested universality of the Method. The meteorological factor c is, for any given locality, peculiar to a certain set of natural conditions pertaining at the given time or over the particular period. It serves as an integrator of these local natural conditions which is not necessarily repetitive. On the other hand, the degree of energy availability $W\varphi$ is in the main unaffected by natural disturbances, has a repetitive or cyclic nature and is solely dependent on the location of the place being considered and on the calendar. The magnitude of the meteorological factor therefore provides the basic measure of the potential rate of evapo-transpiration and the cyclic radiation/latitude factor may be regarded as the 'calibration' factor, both as regards location and time.

The external factors together provide the 'potential' for flow of moisture, *i.e.* loss of water through evapo-transpiration from a crop-soil unit. The form of the equation linking flow and potential tempts one to ask whether 'resistance', analogous to the standard electrical equation, should not be reflected in the function, representing osmotic tension, hydraulic head from root to leaves and friction loss within the soil-plant-air system. The answer is that the resistance element is incorporated in the correlation factor c of the empirical equation, for ideal conditions.

This in turn raises the question of why, by implication, the resistance element is taken to be the same for all crops. The answer must be that this is not the case. Thus the formula does not relate to single plants, but to a crop soil unit involving a population of plants growing in free competition. This is supported by Penman who found that for complete crop covers of different plants having about the same colour and reflectivity, the potential evapo-transpiration rate is the same, irrespective of the plant or soil type.[6] This may also be deduced from the laws of evolution. Plants followed the evolution of soils and adapted themselves to patterns suitable for survival, *i.e.* of highest efficiency and greatest co-operation with soils and environment. This naturally led to selective spacing of different plants within the population to achieve maximum efficiency per unit of area, involving crop and soil. In this connection, it has been noted that correlation of climatic factors is generally better with evapo-transpiration than with evaporation from a tank. This would suggest that the living crop-soil unit, amenable to evolution and adaptation, is more sensitive to climatic impulses than the inert tank units which introduce their own individual characteristics into such response equations.

It is true that an irrigated population is not usually composed of different plants, but in free competition and random spacing the same conditions would generally apply. However, in cases where, for various reasons, there is excessive interference with spacing of plants, the response to climatic stimuli from the crop-soil unit may be affected. This question is referred to later under cropping factors.

It may be possible in due course to write the response equation in terms of pressures and resistances. At present too little is known about the complex constituent resistances which make up the overall resistance to the flow in a crop soil unit.

EVAPORATION IN RELATION TO EVAPO-TRANSPIRATION

In applying such formulae it is useful to know the degree of correlation between potential evapo-transpiration from a crop-soil unit (as forecast) and evaporation from a standard pan. This relationship, which varies with season and latitude, is inherent in equation (1) and may be expressed as follows:

$$E = c/L\varphi \tag{5}$$

in which:

E = free water surface evaporation from a standard tank in mm per day at latitude φ,

C = average depression of the wet bulb in °C for a particular month for a station at latitude φ,

$L\varphi$ = dimensionless cyclic (radiation/latitude) factor for the particular month at latitude φ of the particular locality,

and

$L\varphi = L/Lo$

in which

L and Lo have the meanings previously assigned to them.

From equations (1), (2) and (5) above, it may be seen that

$$Cu\varphi = \frac{E}{L} \tag{6}$$

The use of equation (6) in estimating evapo-transpiration is, however, not generally recommended. As noted by many observers, the rate of evaporation from a can or a pan depends, to a great extent, on the particular characteristics of the receptacle, and varies with pan size, shape, depth, colour, exposure, etc. Moreover, as already mentioned, the correlation between the meteorological factor with evapo-transpiration is better than with free water surface evaporation and thus, except for a cross-check on results, it is generally better to obtain the potential evapo-transpiration curve direct from the meteorological readings.

Perhaps the best known work on the correlation of evapo-transpiration and evapora-tion is that of Penman[6] who, as a result of experimental work at Rothamstead, put forward a suggested method for calculating cropland duties based on free surface evaporation figures, either calculated from his formula or obtained from actual tank observations. It has been found that wherever evaporation readings were reasonably reliable and based on a Standard Weather Bureau Class 'A' or similar tank, the results of the Penman method corresponded closely with those obtained by use of equation (1).

Measurements or estimates of radiation have been used by some authorities to estimate potential evapo-transpiration. Apart from the complexities of measurement and the need for skilled interpretation of results other limitations to this method have been observed in recent papers. These relate to the fact that advective heat, which may under certain conditions (windy and hot days) represent a considerable proportion of the total heat budget, is not taken into account. Here again, it is suggested that the effect of advective heat will be integrated by the readings of the depression of the wet-bulb thermometer and will thus be incorporated in the calculated values of $Cu\varphi$.

In a recent article Jensen and Haise[26] correlate evapo-transpiration and solar radiation for four general climatic regions in the Western United States. A good correlation is observed, but they add: 'The Et/Rs (evapo-transpiration to solar radiation) ratio increases linearly with mean air temperature. There are several reasons for this relationship. As air temperatures increase, the saturated vapour pressure increases in a non-linear manner. Thus, a *larger vapour pressure gradient above* the crop can be expected with higher temperatures, *resulting in more rapid removal of water vapour. . . .*' The $Cu\varphi$ relationship is of course based on the concept expressed in the words which are underlined in the quotation above; c being a measure of the vapour pressure deficit. Jensen and Haise suggest that potential evapo-transpiration may be estimated from mean air temperatures from the equation:

$$Et_p = (0\cdot 14T - 037)\ Rs \qquad (7)$$

in which:

Et_p = potential evapo-transpiration in mm per day,
T = mean air temperature in °C,
Rs = solar radiation in mm per day.

This is directly comparable to the formula $Cu = cW\varphi$ in which:

Cu represents Et_p,
$W\varphi$ represents Rs;
and
c represents $(0\cdot 014T - 0\cdot 37)$.

While Jensen and Haise make use of temperature (T) in their suggested relationship, as does Blaney, it should be noted that they do so because of the relationship between temperature and the saturation vapour pressure.

The Jensen and Haise method involves the measurement or estimation of values of solar radiation (Rs) expressed in mm per day of evaporation equivalent. The methods of estimation referred to by them are in fact somewhat similar in concept to equation (2) above, from which the tabulated values of $W\varphi$ are obtained, except in so far as they suggest, that adjustment should be made for the degree of cloud cover: 'Because solar radiation for a given month and latitude is affected primarily by degree of cloud cover.' While this refinement may be possible in the United States, where fairly adequate records exist, it is doubtful whether it could be made in many other parts of the world where the network of meteorological observing stations is more limited. In equation (1) the cyclic factor $W\varphi$, being based on 'clear-sky' conditions, would not reflect the degree of cloud cover. This is not material because the meteorological factor c does in essence integrate the effect of cloud cover. (The depression of the wet bulb c is in effect a measure of the rate of evaporation of the thin film of water surrounding the bulb, and it will therefore be directly affected by variations in the level of incident radiation caused by cloud cover.)

It appears that relationships linking temperature with consumptive use require the introduction of empirical monthly coefficients which vary both with crop and location, *i.e.* climate.

For instance, in an article[27] on the monthly consumptive use requirements for irrigated crops Blaney tabulates seasonal consumptive use coefficients (K) for various irrigated crops in the Western United States. The consumptive use coefficients for each crop are, however, not constant; he states: 'The lower values of (K) are for coastal areas, the higher values for areas with an arid climate.' It is suggested that this necessity for calibration through the use of differing consumptive use coefficients, may arise from the omission of a latitude factor as well as from the selection of temperature as the basic causative factor.

This leads to a consideration of 'cropping factors' as distinct from 'crop coefficients or factors'. The former relates to an assessment of actual performance in relation to potential consumptive use at a particular time of any crop population located anywhere.

CROPPING FACTORS

By definition[22] [23] $Cu\varphi$ in equation (1) represents: 'The quantity of water, regardless of source, required by a hypothetical crop in full stage of growth in a soil which is being held at a moisture content well above the permanent wilting point.' In other words, $Cu\varphi$ provides an estimate of the *potential* evapo-transpiration from a crop soil unit. The *actual* water loss through evapo-transpiration from a crop-soil unit over a full growing season will, however, vary not only with changing meteorological and energy conditions but also with the stage of development of the crop. Assessment of deviations from potential consumptive use therefore requires the introduction of another variable which may be called a cropping factor. This differs from other methods which incorporate a 'crop factor' as an integral part of the formula, and which have generally been derived from the correlation of observed use of water by specific crops in a particular area with the theoretical use computed by means of a particular formula. In these cases, depending on the extent to which the particular formula completely integrates the various meteorological and other non-crop phenomena, the 'crop factor' so derived may vary from a true 'cropping factor' to a localized calibration factor, covering latitude, energy availability, stage of crop development and other items.

A true crop factor would not vary greatly from place to place, or continent to continent, since it is representative only of the characteristics of the particular plant population and its stage of development. In practice, however, it is found that such crop factors or indices are not necessarily transferable.

It cannot be taken that all crops grown on a commercial basis will, under identical climatic and soil conditions, have the same peak rate of evapo-transpiration. As outlined in the paragraphs below, farming practices such as the wide spacing of trees in orchards for purpose of access or spraying, may have some effect. All that is suggested is that under natural conditions, and given a plentiful supply of water, the response to a given climatic 'pressure' would be the same for most common crops. In the following paragraphs, it is, however, suggested that the effect of plant spacing may be of less importance than other relevant factors.

In the context of this chapter, the term 'cropping factor' may be defined as the ratio between the actual rate of loss of water at any given time from a crop-soil unit through evapo-transpiration and the potential rate of loss of water from the same crop-soil unit, assuming the crop in full stage of growth in the soil which is being held at a moisture content well above the permanent wilting point. In other words:

$$\text{Cropping factor} = \frac{CuA}{Cu\varphi} \qquad (8)$$

In the above equation (8), both the denominator and the numerator are made up of two elements: the evaporation loss from the soil and the transpiration of the plant. The variation of each of these elements with time and in relation to one another is considered briefly below.

Evaporation from a crop-soil unit

Peters,[28] in a recent article, reports, on the basis of field studies comparing the rate of evapo-transpiration from plastic covered and natural plots of corn, that in the Mid-West, where frequent summer showers occur, as much as 50% of the total water loss in a season can be accounted for by evaporation from the soil surface. Some further guide-line on the relative importance of evaporation has also been obtained from data provided in a paper by L. Turc in which data are given on the observed drainage from lysimeters over a period of years (1932–50) at Versailles and elsewhere. At Versailles a series of lysimeters were observed, some with bare soil and some with a variety of crop covers. The data show that, if allowance is made for the period of the cropping season, the average monthly evaporation from bare soil was 30·1 mm/month. This would indicate that, in these experiments, the evaporation loss from a bare soil amounted to 60% of the evapo-transpiration loss from an adjacent crop soil unit. In these tests, the crops grown included barley, wheat, oats, carrots and lucerne. This compares reasonably well with the figures reported by Patric,[29] based on observations carried out at the San Dimas Experimental Station near Los Angeles. Patric's figures, based on lysimetric observations, show that in a dry year, the evaporation from a bare soil may be as much as 47% of the evapo-transpiration from an adjacent grass cover; this figure increased to 52% during a year of above average rainfall.

From the above, it is apparent that when considering the monthly variation in the rate of evapo-transpiration from a crop-soil unit, throughout a season, at least as much consideration should be given to variation in the rate of loss of water through evaporation from the soil as should be given to the variation in the rate of transpiration of the crop.

Variation in the rate of evaporation from the soil

The rate of evaporation from a soil, under a particular set of soil moisture and meteorological conditions, will be dependent on the net radiation at the soil surface. The energy absorbed will be used in heating the soil and in the evaporation process. It is difficult to determine the proportion of the energy used to heat the soil, but several researchers have indicated that over short periods of time the magnitude of soil heat, with respect to the other processes, is small.

The development of a crop will reduce the amount of solar radiation reaching the soil surface, and thereby reduce the net radiation at the soil surface and hence the rate of evaporation. The extent to which the crop will affect the rate of evaporation by absorption of incoming solar radiation will vary with the denseness of the cover provided. In this connection, the stage of development of the crop will be of prime importance.

Shaw[30] has reported that, based on radiation measurements in a corn crop in Iowa, the amount of energy available for evaporation from the ground surface in a corn crop is close to 100% of the net, early in the season. When the crop had reached a height of about 1·5 m (about two-thirds of its final height), there was still 60–65% as much net radiation above the ground in the crop cover as there was above the crop. When the corn reached its maximum height, it was estimated that the figure had been reduced to 14%

In the foregoing, consideration of the variation in the rate of evaporation is limited to the phase from planting or emergence to the stage of full growth, which is the most important period for most crops. However, if the analysis is extended to cover the full natural cycle of a plant's life, the stage beyond maturity to fruiting and eventual desiccation should also be taken into account. During this second stage, the interception of radiation by the plant's foliage will gradually decrease, and the potential for evaporation of water from the soil will increase.

It is therefore suggested that under constant meteorological conditions and providing adequate moisture is available at all times, the rate of loss of water, over a growing season, by evaporation of water from the soil in a crop-soil unit may be represented diagrammatically by a curve as shown in Fig. 11.

Transpiration of the crop

At leaf emergence, due to the lack of transpiring media, the crop will transpire at a comparatively slow rate, and the rate of increase will be comparatively slow, as the root structure develops.

During the vegetative growth stage, which precedes flowering when the root structure is well developed, the foliage will develop rapidly with a consequent fairly rapid increase

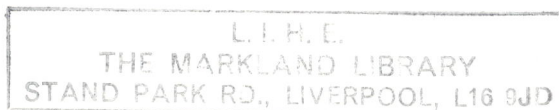

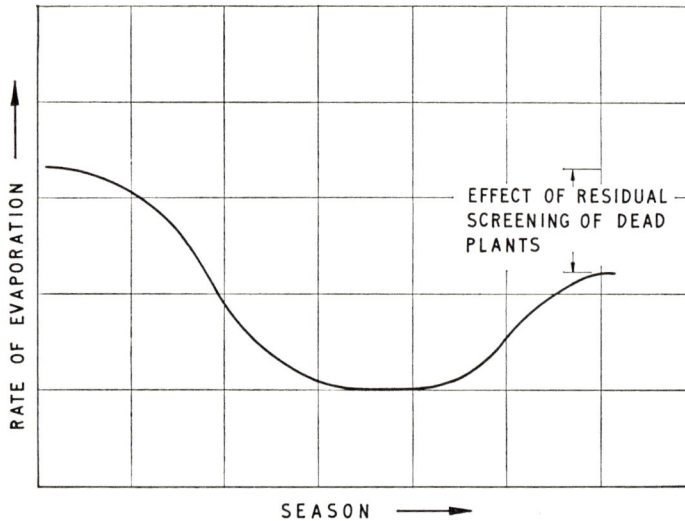

Fig. 11 Diagrammatic representation of seasonal variation in rate of evaporation under constant
     meteorological conditions and given a full supply of moisture in all times.

in the transpiration rate of the plant. The peak rate of transpiration will normally be
reached at, or near, the flowering stage, when the plant is mature, and will generally last
throughout the fruit formation stage. If the crop is of the type that is harvested at the 'dry
fruit' stage, the crop will remain in the ground, the foliage will begin to fall or wither, and
the rate of transpiration will decrease until the plants are dead, and transpiration will
cease altogether.

Given constant meteorological conditions and free availability of water to the plants,
it is therefore suggested that the rate of loss of water from a crop-soil unit, due to
transpiration, could be represented diagrammatically by a curve as shown in Fig. 12.

Variation in combined rates of evaporation and transpiration

The diagrammatic variation of potential evaporation and potential transpiration, shown
in Figs. 11 and 12, if combined, would tend to show that under constant meteorological
conditions, and given a continuous free supply of moisture, the evapo-transpiration from
a crop-soil unit, over a full season, could be represented diagrammatically as shown in
Fig. 13.

In the preceding paragraphs, the water loss through evapo-transpiration, under
constant meteorological conditions, has been considered in general terms assuming that
the soil moisture content is at all times kept at a high level, so that evapo-transpiration is
not inhibited due to lack of moisture, and the effect, if any, of the spacing of plants and
plant population has been ignored.

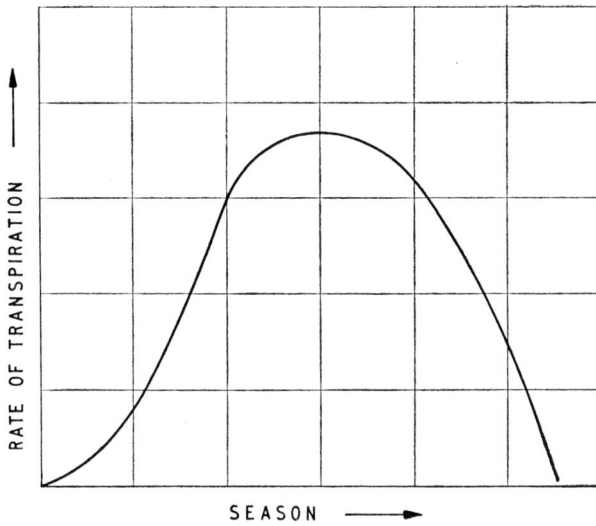

Fig. 12 Diagrammatic representation of seasonal variation in rate of transpiration under constant meteorological conditions.

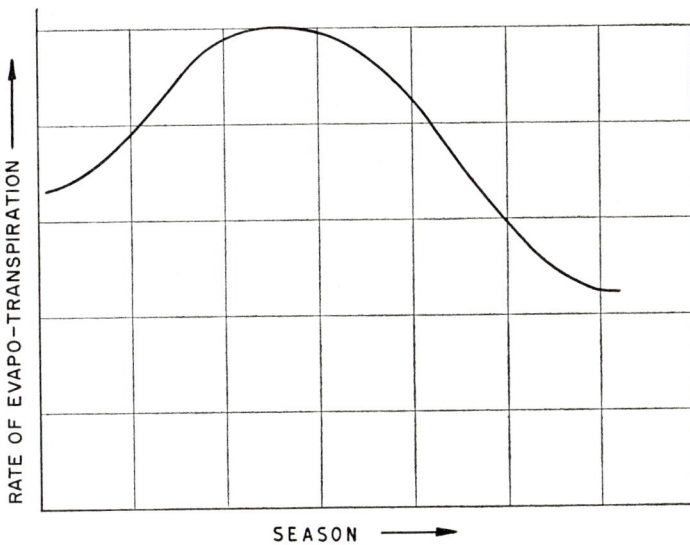

Fig. 13 Diagrammatic representation of seasonal variation in rate of evapo-transpiration under constant meteorological conditions and given a full supply of moisture at all times.

A reduction in soil moisture will, however, reduce the rate of evapo-transpiration. As reported by Bahrani and Taylor,[31] based on field experiments with lucerne (alfalfa), the amount of water used by the crop-soil unit decreased with increase in the mean integrated moisture potential in the root zone, and the actual evapo-transpiration and its ratio to 'potential evapo-transpiration' (calculated by Penman formula) showed a decreasing curvilinear relationship with the average moisture potential.

Willardson and Pope have also shown that, based on an analysis of data presented by Wilcox concerning evapo-transpiration from lucerne and grass, the rate of evapo-transpiration may fall by as much as 50% over a period of 20 days after an application of irrigation water.

This phenomenon is probably due more to a reduction in the rate of evaporation than to a fall-off in the rate of transpiration. Following an application of irrigation or of rain the degree of radiation absorption will decrease as water is lost from the soil, *i.e.* as the soil profile dries out, and consequently the *actual rate* of evaporation will drop below the *potential rate* of evaporation. The plant, with its rooting system well below the soil surface, is, however, protected against the immediate effects of the drying out process. Also, providing the soil moisture content in the root zone does not drop below the wilting point, the rate of transpiration by the crop should remain fairly constant. This is in accordance with the findings of Veihmeyer and Hendrickson. Also, Gardner and Ehlig,[32] based on the results of transpiration measurements of pepper plants and birds-foot trefoil grown in shallow glazed crocks in a greenhouse, found that there was 'little variation in the transpiration rate (with decreasing average soil-water content) until the plants wilt; thereafter, there is virtually a linear relationship between water content and transpiration rate'.

In irrigated agriculture the effect of drying of the soil will be particularly felt towards the end of the season. At this time, with the root structure fully developed, the plant may draw its necessary supplies of water from a considerable depth in the soil profile, irrigation may be withheld (providing the moisture content of the soil at depth has been built up over the season) and evaporation may represent a negligible proportion of the total evapo-transpiration. Early in the season, however, before the rooting system has developed, it will normally be desirable to maintain a fairly high moisture content in the upper section of the soil profile; actual loss of water through evaporation may be important during this period.

These considerations apply, of course, only to dry foot crops; in the case of wet foot crops such as rice, evaporation will represent a greater, and more constant, proportion of evaporation.

Figure 14 represents, diagrammatically, the seasonal variation of the ratio of actual evapo-transpiration to potential evapo-transpiration for a crop-soil unit in which the plants are allowed to remain in the ground until they are completely wilted. Many crops are, however, harvested, and the field cleared, before the full natural cycle of the plants' life is completed. In this case, the cycle will be cut short, and the seasonal variation of the cropping factor will terminate at some point, A or B.

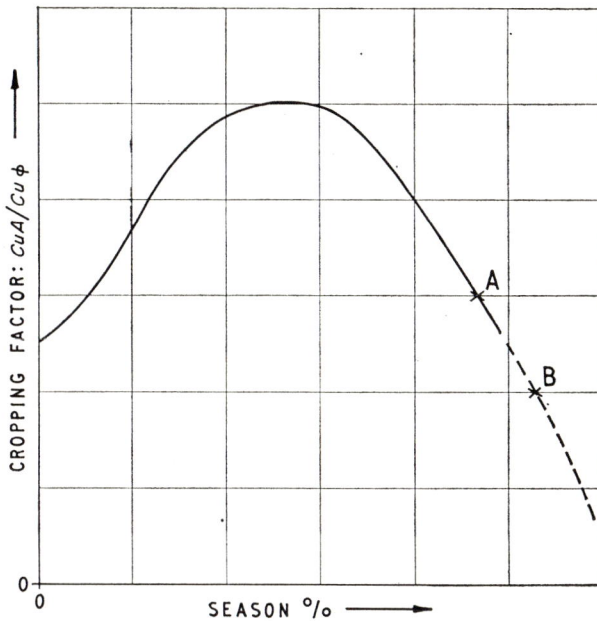

Fig. 14 Suggested seasonal variation of cropping factor: $CuA/Cu\varphi$.

GENERAL APPLICATION

The engineer's functions with regard to development of water resources are to assess
storage requirements, and to design and construct dams, weirs, conduits and channels by
which water is delivered to the land. Normally his interest should not extend across the
boundaries of, or on to, the fields. For his purpose it is necessary to be supplied with the
following information which comes rightly within the province of the agriculturist with
whom the closest liaison must be maintained at all stages of the project:

  (i) The amounts of water required on the land;
  (ii) The times and durations for which such water is needed.

   Agricultural advice is not always available, especially in the early stages of some
feasibility studies, and in these circumstances the engineer should be able to form his
own assessment of requirements. Moreover, the need to comply with economic criteria
and time disciplines and to make full use of the advantages of aids from the computer
makes it essential for all studies to commence *at the grass roots, i.e.* in the fields and to
work upstream towards storage and diversion structures.
   Subject to the adjustments for cropping factors referred to it is now possible to use

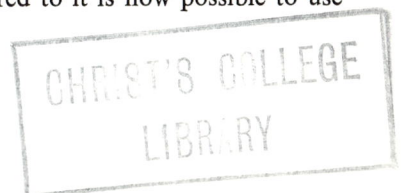

reliable formulae to determine the basic irrigation characteristic for any given locality. Basic irrigation characteristic may be defined as the theoretical water requirements curve adjusted for effective precipitation and before adjustments are made for cropping factors and patterns.

In drawing up such characteristic curves it is perhaps advisable to use one or two formulae as a cross-check on each other, as is illustrated in Table 6,[33] which relates to a recent report for the Government of Kuwait.

Table 6
COMPUTED POTENTIAL EVAPO-TRANSPIRATION
(mm/month)

Month	Olivier method		Penman method	
	Khuzestan	Kuwait	Khuzestan	Kuwait
Jan.	22	35	60	51
Feb.	37	58	71	73
Mar.	98	114	119	129
Apr.	180	171	203	177
May	259	267	257	264
Jun.	338	354	297	315
Jul.	344	361	316	321
Aug.	295	309	267	280
Sep.	222	249	202	212
Oct.	130	151	148	156
Nov.	47	64	82	80
Dec.	17	31	56	49
Annual total	1989	2164	2078	2107

It cannot be stressed strongly enough that no formula, complex or otherwise, can ever provide a universal or 'push button' solution: the onus will always rest with the engineer and agricultural officers to exercise experienced judgements concerning special factors whether pertaining to physical conditions or to modes of practice.

The following examples, though by no means exhaustive, serve to illustrate to some extent what is meant:

(i) Our knowledge of what is really effective rainfall in various parts of the world is still pitifully scant. In these circumstances careful judgement must be used by the calculator, based on practical experience, in arriving at the modifications to be made to the basic characteristic curves.

(ii) The amount of water indicated as basic irrigation requirement for a particular application may be below that which, in the absence of special equipment, can be supplied by furrow irrigation methods. It is a difficult matter to apply by furrow systems a uniform depth of less than twenty-five millimetres per watering, for instance. Here again judgement will be required to assess the actual practical as compared with theoretical forecast requirements.

(iii) Where fine irrigation spray systems are used it must be borne in mind that water is 'lost', *i.e.* not available to the root zones due to the fact that a proportion of the fine spray drifts or blows away ineffectively and also due to evaporation from leaf interception. It will therefore be necessary in practice to allow for additional water above the theoretical (indicated) requirements.

It is emphasized that whereas the use of these formulae does enable engineers to make rapid estimates of basic field water requirements for new schemes, it must not be overlooked that the resulting characteristics are based on the assumption that, for these new areas, the soils and other factors are suitable for the particular crops on a long-term basis. Therefore the actual planning and implementation of such schemes should in all cases be worked out in close consultation with competent agriculturists and agronomists.

The usefulness of reliable estimates of crop-soil water requirements is, however, not limited to the overall planning and operation of an irrigation system. It has been demonstrated that variations in irrigation water costs and in quantities available (or used) brings changes in crop organization and in net returns.[34] [35]

Most farm operators on occasions find it necessary under scarcity conditions to ration the limited quantities of irrigation water available for crop use. Some face this problem practically every season; others only under unusual conditions, such as reduced supplies accompanying years of low precipitation. Regardless of when a farmer must allocate water among competing farm uses, it is highly important that he have effective guides for making sound decisions. He needs reliable information that will enable him to allot his scarce supplies so as to maximize his earnings.

Through a linear programming analysis researchers[36] in a recent study sought to establish changes in total net farm returns and the dollar value increments (marginal value products) involved in water changes for a given cropping pattern. It was found that crops vary widely in net returns to added quantities of irrigation water as illustrated in Table 7.

The authors concluded as follows:

'Clearly cotton offers the highest net return for the first irrigation water and should receive all that can be obtained up to 900 acre-feet (1.11×10^6 m^3). Black beans, and then alfalfa hay, on Grade I soil as is the cotton, earns the greatest number of dollars per acre-foot for additional water up to about 1150 acre-feet (1.42×10^6 m^3) total for the farm. Finally, if still more water can be obtained, alfalfa hay on the Grade II soil (181 acres) will return $13.34 over the $3.00 price used in this analysis, or a total of $16.34 per acre-foot ($13.1 per 1000m^3) for an additional 1146 acre-feet (1.41×10^6m^3), making a total quantity of 2762 acre-feet (3.41×10^6 m^3) for the System C farm. Marginal value products (dollars added per acre-foot of irrigation water) decline sharply from $59.00 per acre-foot ($47.80 per 1000 m^3) for the first 900 acre-feet used on cotton to $16.00 per acre-foot ($12.96 per 1000 m^3) for the final 1150 acre-foot increment applied to alfalfa hay on the Grade II soil.

Table 7

COMPARISON OF NET RETURNS AND MARGINAL VALUE OF PRODUCTS WITH INCREASES IN IRRIGATION WATER SUPPLIED TO A GIVEN CROPPING PATTERN (also see Table 5(a) in the Appendix)

Water		Net returns		Marginal value of products per added 1000 m³		Crops	
10⁶m³		Dollars		Dollars		Name	Hectares
Added	Total	Added	Total	Water @ $2·43	Water @ $0·00		
1·103	1·103	49 862	49 862	45·22	47·62	Cotton	81
0·425	1·528	13 981	63 843	32·90	35·30	Cotton B. beans Total	81 65 146
0·463	1·991	5 025	68 868	10·85	13·28	Cotton B. beans Alfalfa Total	81 65 25 171
1·414	3·405	15 285	85 152	10·84	13·27	Same Alfalfa Total	171 73 244

'If an operator under these circumstances has less than the 900 acre-feet ($1·11 \times 10^6$ m³) of water required for irrigating all 200 acres (81 hectares) of cotton properly, he should reduce acreage accordingly. He definitely should not attempt to ration the water among the total 200 acres that would maximize net returns under more favourable water conditions. He would find it more profitable to leave acres idle than to attempt to spread inadequate quantities of irrigation water over too many acres.'

The results of a separate linear programming analysis to determine changes in total farm net returns over variable expenses for each of three cropping systems (A, B and C) with variations in irrigation water costs from zero to approximately U.S. $3.00 per acre-foot ($2.43 per 1000 m³) are summarized in Fig. 15.

In this exercise irrigation water variable expenses (costs) include only direct cash operating costs for pumped water and tolls paid for surface water. Fixed overhead costs such as depreciation, taxes and other items for pumped water and assessments for surface water are part of the total 640 acres (263 hectares) farm fixed costs, taken as U.S. $64 000 for the study.

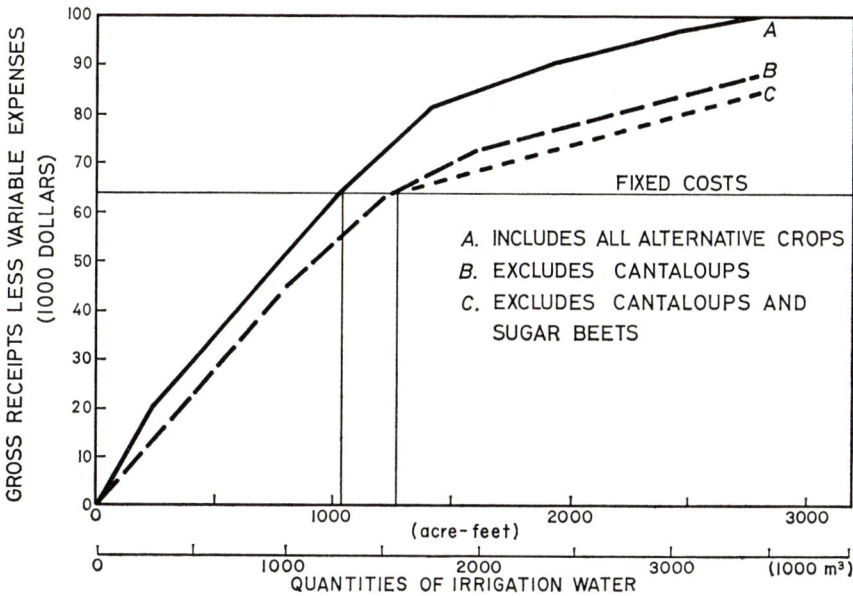

Fig. 15 Farm net returns at varying quantities of irrigation water; three cropping systems.

APPENDIX TO CHAPTER 2
MONTHLY FIELD WATER REQUIREMENT CHARACTERISTIC CONSTANTS FOR
DIFFERENT LATITUDES: $W\theta N$

	$W\theta N$ for different degrees of latitude North											
Month	0°	5°	10°	15°	20°	25°	30°	35°	40°	45°	50°	55°
Jan.	0·64	0·61	0·56	0·52	0·47	0·42	0·38	0·32	0·26	0·21	0·16	0·11
Feb.	0·69	0·66	0·64	0·60	0·56	0·50	0·46	0·42	0·37	0·32	0·27	0·22
Mar.	0·74	0·73	0·71	0·70	0·67	0·64	0·62	0·60	0·56	0·52	0·49	0·46
Apr.	0·72	0·74	0·74	0·75	0·75	0·75	0·75	0·73	0·73	0·73	0·72	0·73
May	0·64	0·69	0·73	0·74	0·77	0·79	0·80	0·83	0·83	0·84	0·87	0·91
Jun.	0·61	0·65	0·70	0·74	0·76	0·79	0·83	0·83	0·86	0·89	0·91	0·96
Jul.	0·62	0·67	0·70	0·74	0·76	0·79	0·82	0·83	0·85	0·86	0·89	0·95
Aug.	0·69	0·71	0·72	0·72	0·73	0·72	0·73	0·70	0·71	0·71	0·70	0·71
Sep.	0·73	0·73	0·74	0·73	0·72	0·70	·068	0·67	0·66	0·63	0·60	0·61
Oct.	0·72	0·69	0·66	0·64	0·62	0·60	0·54	0·49	0·46	0·41	0·37	0·33
Nov.	0·66	0·61	0·58	0·54	0·50	0·45	0·39	0·34	0·29	0·24	0·19	0·14
Dec.	0·61	0·57	0·53	0·47	0·43	0·37	0·32	0·27	9·21	0·17	0·11	0·07

3 Time disciplines in water resources planning

It is not easy to define in simple terms time disciplines in hydraulic engineering. We are only now becoming conscious of the practical and far-reaching implications of the fourth dimension, time, in all aspects of water resources planning, on both sides of the ledger: requirements (demand) and provision (supply). Indeed, it is not always easy to draw a clear distinction between supply and demand factors.

Disciplines imposed by *requirements*, *i.e.* the factors which create the need for the engineering structures, are not as rigidly fixed as those which relate to the media associated with the dynamics of water. Thus, whereas it may be possible to retard or halt the population explosion, or alter the times when television programmes are switched off and tea kettles are switched on, thereby causing heavy electricity peak loads requiring sudden water releases through turbines, it is not foreseeably easy to alter and predict the incidence and rates of rainfall, the rate of scour and degradation of rivers, and changes in characteristics of catchment areas.

A useful purpose might be served if some relevant issues are highlighted by selection of a few specific examples from the demand and supply categories.

REQUIREMENTS (DEMAND)

Undoubtedly the greatest time factor affecting requirements stems from the rate of increase in population and from increasing standards of living.

The growth of water use

Up to the nineteenth century water use was confined primarily to agricultural and domestic use. These developments waxed and waned in relatively localized communities because, in the era of camel communications, neither qualitative nor quantitative appreciation of the benefits of irrigation was easy to disseminate. It is very likely that degeneration of the vast and sophisticated irrigation systems of the Lower Euphrates Valley (traces of which remain to this day) was due to administrative neglect as well as progressive siltation and salinization.

As mentioned in Chapter 1, the changes of the Industrial Revolution of Europe caused a rapid redistribution of populations and consequently a realization of the term 'standard of living'. The effects of recurring famines on populations were observed, reported and interpreted. There also dawned an increasing awareness of the fact that shrinking agricultural communities would need to support increasing non-agricultural communities, thereby incurring expanding obligations over and above those of self-sufficiency. This trend has continued to this day as may be seen from Table 8.[37]

Table 8
NUMBER OF PERSONS
SUPPLIED BY ONE FARM
WORKER IN THE UNITED
STATES

Year	Number of persons supplied
1850	4·18
1860	4·53
1870	5·14
1880	5·57
1890	5·77
1900	6·95
1910	7·07
1920	8·27
1930	9·75
1940	10·69
1950	14·56
1960	26·34

It has been estimated that during the nineteenth century the world's irrigated areas increased from 8 million to 40 million hectares (20 million to 100 million acres).

In the first half of this century the total area again increased rapidly to over 160 million hectares (about 400 million acres) in a time when the world population increased from 1600 million to 2500 million, at a rate varying between 0·50% and 2·5% per annum for different countries. The average rate of increase of irrigated area was about 2·4 million hectares (about 6 million acres) per annum.

The estimated total areas under irrigation for the principal irrigating countries, for the period 1957–61, are set out in Table 9.

It will be seen that China, India and Pakistan now account for approximately two-thirds of the world's irrigated area.

As stated in Chapter 1, it has recently become apparent that apart from considerations of economic viability in terms of monetary values, which are subject to inflation or devaluation, there is the question of nutritive viability, agricultural yields in calories per unit of land. The force of this point may be illustrated by reference to Table 10 which compares results of recent surveys for selected countries.[10]

Taking mimimum adequate standards as 2500 kCal/head, it is apparent from these figures that if the task of keeping the U.S.A. a 'food adequate' area is considered by some to be difficult, then how immensely pressing and difficult will be the tasks of providing, by the year 2000, the additional irrigated area for India, and China, in order to maintain (let alone improve) diet levels which are already below minimum standard.

Over the years, however, increased research activity covering all fields and the developments of chemical fertilizers, pesticides, improved seeds and herbicides helped to lift agricultural yields to increasingly high ceilings. A striking example of the pattern of

Table 9
PRINCIPAL IRRIGATION AREAS OF THE WORLD[9]
(Land receiving water by irrigation schemes excluding inundation areas)
(also see Table 6(a) in the Appendix)

Order	Country	km² × 10³	Order	Country	km² × 10³
1	China	740·0	16	Chile	14·13
2	India	234·5	17	Peru	12·12
3	U.S.A.	152·4	18	Korea	12·12
4	Pakistan	111·0	19	Australia	8·50
5	Russia	71·9	20	Philippines	8·09
6	Indonesia	59·8	21	Sudan	8·09
7	Iran	46·8	22	Madagascar	7·29
8	Mexico	42·0	23	Viet Nam	6·07
9	Iraq	36·8	24	South Africa	6·07
10	Japan	34·0	25	Syria	5·66
11	Egypt	24·6	26	Morocco	5·26
12	Turkey	19·8	27	Burma	5·26
13	Spain	18·6	28	Colombia	4·85
14	Thailand	16·6	29	Formosa	4·85
15	Argentina	15·0	30	Greece	4·45

Table 10
DIET LEVELS IN RELATION TO EXISTING AND PROJECTED IRRIGATED AREAS
(also see Table 7(a) in the Appendix)

Country and projection period		Population (× 10⁶)	National mean daily diet levels (kCal/head)	Total area irrigated km² × 10³
Egypt	1961	26·6	2530	24·6
	2000	56·0	1800	34·0
Sudan	1960	12·1	2500	8·1
	2000	25·0	2500	18·2
India	1960	442·0	2040	234·5*
	2000	670·0	2300	441·0*
West Pakistan	1960	43·0	1970	68·8†
	2000	65·0	2300	129·2†
U.S.A.	1960	184	3100	165·8
	2000	311†	3100‡	1132·0
Mainland China	1960	640	1900	740·0
	2000	1380	2000	1830·0

Notes: * Includes a considerable proportion of double-cropped km²
† Includes a considerable proportion of double-cropped km² and relates only to areas irrigated by perennial canals
‡ High projection

advances achieved over a long period in yields per unit area may be found by examination of the historical records of two staple crops: wheat in England and rice in Japan. Wheat yields in England increased slowly from approximately 335 kg/ha (about 300 lb/acre) around A.D. 1250 to approximately 2240 kg/ha (2000 lb/acre) by 1900 and then very rapidly to 3350 kg/ha (3000 lb/acre) by 1950. Rice yields in Japan followed a similar pattern: a slow increase from 1000 kg/ha to 2000 kg/ha over the period A.D. 1200–1900 and then a meteoric increase over 50 years up to over 4500 kg/ha (4000 lb/acre), Fig. 16.[38]

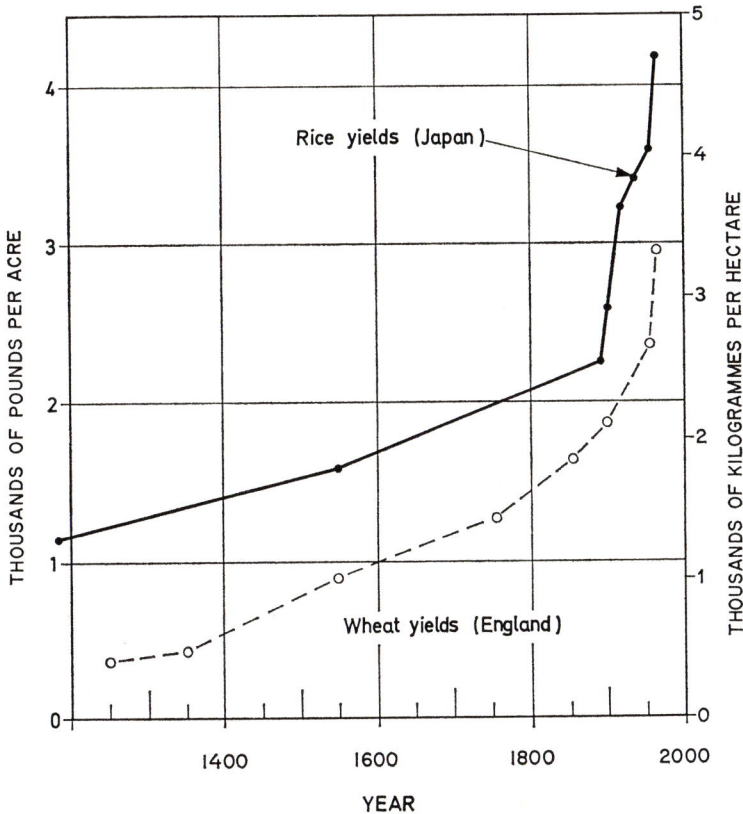

Fig. 16 Historical trends in rice and wheat yields.

It is important to realize that the dynamic trends are not of interest merely because of the rate of *total* population increase, but because of the *relative rates of increase between the more and the less advanced regions.* Reference is made in Chapter 5 to the fact that, according to recent World Bank Surveys, between the years 1800 and 1964 the world population rose from 923 million to 3286 million, of which only 900 million were located in regions of advanced development. This gap is increasing rapidly.

Inputs other than water

It must be borne in mind that the efficiency of irrigated agriculture does not depend only on water as an input. In many areas institutional constraints, lack of education and extension services cause undue delays with respect to application of modern agricultural techniques. Because the provision of additional water supplies is not so restricted and agricultural benefit can be obtained in a relatively short period, there is frequently a tendency, under pressure, to choose the water input to provide immediate impetus to agricultural yield. This may not necessarily be wrong, but when applied without proper consideration of all the factors there is scope for serious mistakes with consequent long-term effects such as raised water tables and increase in salinity of soils.

The expenditure of effort and money on engineering and administrative improvements as regards water application must be accompanied by other equally important improvements directly concerned with water use for crops. These include cultural practices such as pest and weed control, preplanting cultivation and the application of fertilizers, aimed at increasing the depth and nutrient status of the root zone, thus increasing crop yields and effective soil moisture storage. Advances in plant breeding directed towards improved yields and rooting habits under irrigated environmental conditions could also contribute greatly in this respect.

The time element arises from the question as to how long a particular community will take to implement improved practices.

In planning a project analysis in a developing country, forward projections are frequently done on a three-tier basis:

Low input level (pessimistic):
Apart from full water supplies it is assumed that during the projection period improvements in production will be effected only slowly and that such improvements would result from gradual introduction of better ploughs and harrows and a limited introduction of mechanization and use of fertilizers. It is assumed that constraints as regards credit facilities and land tenure will not alter to a great extent.

Medium input level (realistic):
It is assumed that there will be gradual, but steady, modernization in all fields and that, at the end of the projection period, only the most tenacious constraints, such as land tenure, size of holdings, etc., will remain.

High input level (optimistic):
Projection is based on the assumption that all constraints are removed during the period under consideration: fertilizer is used in adequate quantities, mechanization has been fully accomplished, farm management and extension services are adequate, credit facilities are available and size of holdings is determined by sound economic/financial criteria.

A recent survey[39] carried out in West Pakistan for a 50-year projection period (1974–2024) indicated yield projections for three crops as shown in Table 11.

Table 11
SUMMARY YIELD PROJECTION FOR DIFFERENT INPUT LEVELS
(kg/ha × 100)
(also see Table 8(a) in the Appendix)

Crop	Cotton			Fodder			Wheat		
Input level	L	M	H	L	M	H	L	M	H
Present yield	7·15	7·15	7·15	237	237	237	11·25	11·25	11·25
1974 yield	9·95	8·95	8·95	302	302	302	13·30	13·30	13·30
2024 yield	14·30	27·50	31·10	456	705	769	18·75	36·50	39·00
Average yearly rate of increase in yield 1974/2024	1·2%	4·1%	4·8%	1·0%	2·6%	3·1%	0·8%	3·5%	3·9%

Notes: L: Low input level
M: Medium input level
H: High input level

Changes in demand

Industrial expansion in the past 50 years has been accompanied by a steady increase in water requirements so that agricultural requirements have, in many instances, been surpassed and in some basins grave shortages are threatened.

Table 12 shows that in the U.S.A. by 1955 the general industrial requirements amounted to half the irrigation requirements, and the total industrial (including power and municipal) requirements came to 50% of total usage.

However, it is also noted that projected usages reveal a reversal of trend and that agricultural requirements, assuming current practices, would exceed industrial needs between 1975 and 1995. As noted in Chapter 1, however, it is proposed to double the 1965 irrigation area by the year 2000 with hardly any change in water requirements by achieving vigorous—and ruthless—improvements in efficiency of agricultural usage. Also the projection of industrial usage envisages considerable improvement in the industrial use of water. This involves time disciplines with particular reference to re-use of water.

The projected increase in water usage for electric power production following from a rising standard of living and industrial development is also clearly shown in Table 12.

In recent years the time disciplines of the hydraulic and the electrical engineer have become inextricably interlinked. As a result of technological advances both in electrical generation and long distance transmission coupled with changing patterns of public demand for electric energy the paradoxical position has been reached where the maximum economic benefit from storage projects may only be reached towards the end of the 'life' of the structure. This has led most countries to review their outlook on hydro-electric potential. One aspect, namely pumped storage, has been selected to illustrate the implications of this trend.

The essential feature of pumped storage is to store energy during periods when efficient

Table 12
ESTIMATED FUTURE WATER USE IN THE U.S.A.[40]
(also see Table 9(a) in the Appendix)

| Year | Water uses $m^3 \times 10^9$ | | | | | Percent of total supply |
	Irrigation	Domestic and municipal	Electric power	General industry	Total	
1955	112·0	14·9	82·7	56·1	265·7	16·3
1965	116·6	19·7	102·3	67·8	306·4	18·8
1975	135·8	24·7	123·2	83·9	367·6	22·6
1995	384·0	34·6	179·0	113·5	711·1	43·7

generating capacity exceeds consumers' demands and to release this energy whenever the demand on the system exceeds the output which can be rapidly or efficiently mobilized by the principal base load stations. The earlier installations were based on the concept of running thermal installations at constant load and therefore at optimum efficiency. At nights and weekends their outputs therefore exceeded demand and the excess energy thus made available was used to pump water to a high level where it was available for release through a conventional turbine during subsequent peaks.

In recent years, however, pumped storage has taken on a new importance with the introduction of nuclear power stations and with the tendency for conventional steam sets to increase in size. This increase in size has rendered the thermal installations less capable of responding to rapid changes in load and pumped storage plant has therefore tended to become a necessity rather than merely an advantage.

Along with these changes in the thermal sector of power generation, there have been significant advances in design of hydraulic turbines and equipment for pumped storage installations. The reversible pump turbine is now widely used and has been built to a maximum capacity of 100 MW; reversible pump turbines of 300 MW are under construction and even greater unit capacity is now being seriously contemplated.

Recent experience indicates that the traditional daily load pattern is altering. In modern conditions rapid fluctuations can occur, such for instance as the 'television peak' which occurs at the end of a popular programme when several million people switch off their television sets and switch on all the lights and their electric kettles. To meet such new conditions pumped storage blocks are being designed for increasingly more rapid rates of pick-up. The Ffestiniog scheme has been brought up to full load in 51 seconds and even more rapid rates of loading are contemplated for future installations.

Paradoxically, the outstanding feature of a pumped storage scheme is that in comparison with conventional hydro it requires very little storage, the reason being that it is only storing energy for a few hours instead of having to span an annual hydrological cycle for several successive dry years. This points to the possibility of developing conventional hydro stations for conversion to pumped storage when their storage capacity has been reduced by siltation.

Progress in the field of water treatment may also considerably affect future water use. Domestic water is not necessarily wholly consumptive since a large proportion of the water is used for washing and sanitation purposes, both of which are in themselves non-consumptive.

A recent survey[41] for the City of Johannesburg in South Africa illustrates a surprisingly low figure for consumptive use, owing to the high degree of purification achieved after primary use.

Table 13
ESTIMATED WATER UTILIZATION: JOHANNESBURG 1965/66

Use	Proportion of total supply	Ratio of non-consumptive to consumptive use	Actual consumptive use
Domestic	61%	85 : 15	9·1%
Commerce and industry	17%	95 : 5	0·9%
Municipal, provincial and government	22%	75 : 25	5·5%
	100%		15·5%

In some countries, such as Israel, regulations are already in force with regard to the use of purified effluent for the irrigation of the following restricted crops:

Industrial crops, e.g. cotton.
Pasture, provided the animals start grazing when the grass is dry.
Grass grown for hay.
Vegetables which are only consumed after cooking, e.g. potato and eggplant.
Citrus, bananas, nut trees, date palms and avocado pears.
Ornamental shrubs, plants and flowers.
Plants grown for seeds.
Sunflowers and carobs, provided irrigation is by furrow only.
Apple, pear, peach and plum trees, provided irrigation is stopped at least one month before harvesting.

PROVISION (SUPPLY)

Irrigation demand characteristics
Three sets of hydraulic laws, all subject to the discipline of time, govern irrigation demand characteristics:

(i) Storage and conveyance to the fields.
(ii) Charging the root zone.
(iii) Crop-soil consumption, i.e. return of water to atmosphere.

The first two are dealt with in greater detail in Chapter 7, while the basic principles of crop-soil water use have been referred to in Chapter 2. Some of the time factors inherent in these laws are, however, referred to below.

Some typical characteristics for basic water requirements for selected countries in relation to time are illustrated in Fig. 17. It will be seen that these characteristics which represent potential requirements according to climatic and latitude factors are generally similar in shape but vary in amplitude.

To derive the corresponding irrigation requirements such curves have to be adjusted for effects of other factors which are themselves subject to time disciplines.

(a) *Effective rainfall*
In an arid country, *i.e.* where there is no effective rainfall at any time of the growing season, such as Egypt, basic characteristics will also be the irrigation characteristic. In countries where the effective seasonal rainfall distribution matches the basic water requirements characteristic no irrigation will be required. In between lies an infinite variety of rainfall conditions which, according to its 'effectiveness', particularly as regards distribution in relation to growing periods, require different rates and amounts of irrigation. This procedure of making up deficiencies in effective rainfall has become known as *supplementary* irrigation. Obviously efficiency depends to a large extent on considerations of timing and rates. Inadequacies of current methods of measuring rain-fall, particularly rates of incidence, is a great handicap.

(b) *Soil type—infiltration rate*
Reference has already been made to the rate of infiltration of applied water. At one end of the scale the heavy clays absorb a relatively low rate of supply and at the other extreme very sandy soils absorb relatively fast rates. Thus, for sandy soils an area supplied by furrow at too low a rate will suffer over-watering (*i.e.* waste) in the upper section of the area and under-watering (*i.e.* starvation) in lower sections. The reverse applies to a clay soil fed at too high a rate.

(c) *Root zone storage*
Soil storage for irrigation purposes is a function of soil voids and plant root depths. The usable moisture is the quantity available in the range lying between soil field capacity and wilting point. The deeper rooted plants in any soil, owing to the greater storage capacity from which they may draw water, will provide an in-built regulating reservoir to the crop. Also for a given rain 'hydrograph' (or hyetograph) the crop-soil unit makes more effective use of the total incidence than the equivalent bare-soil unit. Interception by leaves and stems regulates the rainfall intensity at soil surface level in exactly the same manner as a reservoir attenuates the peak of an incoming flood. Thus, a rainfall intensity on a clay soil which would have been quite ineffective is made effective because of increased potential infiltration caused by redistribution of incidence rates. The operating concept from the irrigator's point of view is similar to that of the operator of river storage projects: there are two potential 'reservoirs' in series which can be operated to maximize

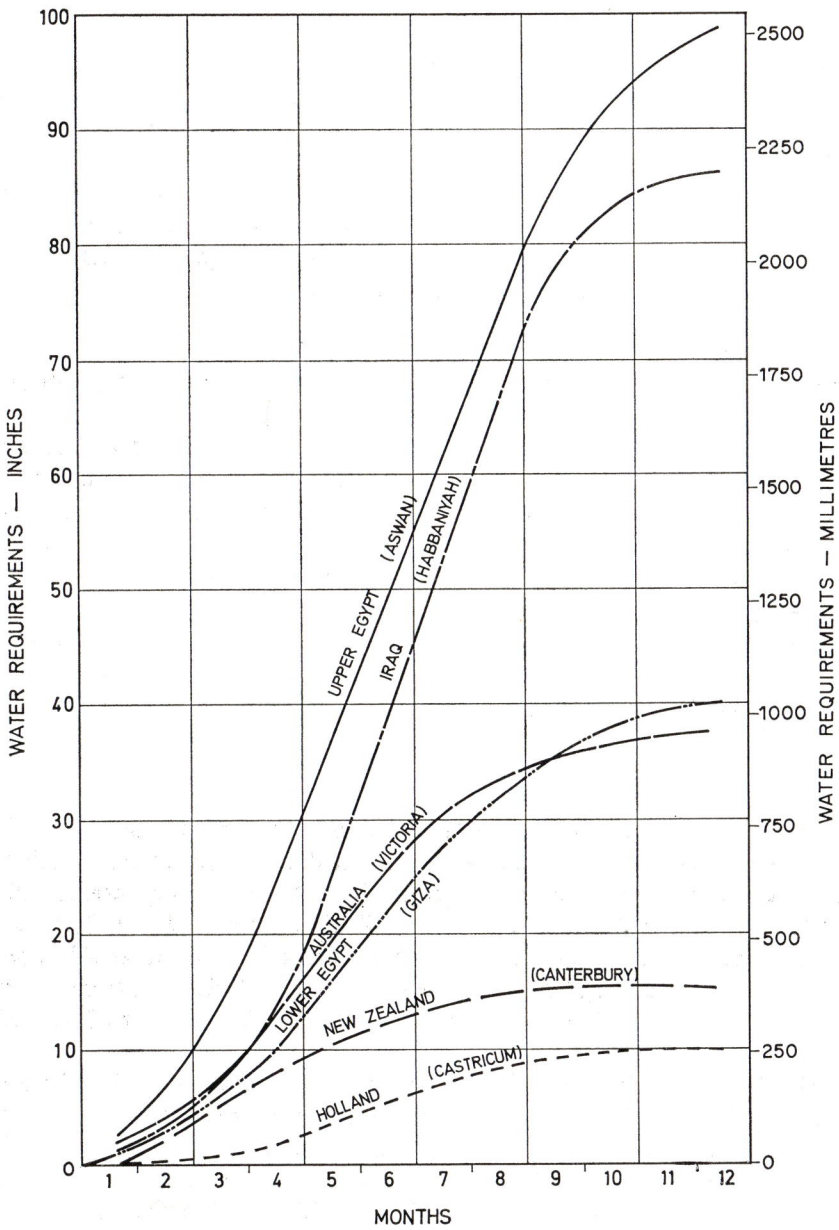

Fig. 17 Typical average annual characteristic cropland water requirement curves (*Cu*).

c

the benefits per unit of water. Two deductions flow immediately from this thought stream.

(i) It should be possible to maximize benefits from the primary reservoir (rain or spray incidence) for a given climatic time pattern by suitable disposition of intercepting cover, whether by crops or by artificial mulches. This may have the effect of reducing fallows.

(ii) The potential of the secondary reservoir—soil storage—may be much improved by development of deeper rooted crops. A corollary concept concerns the use of deep rooted crops or phreatophytes to draw down high soil moisture levels to alleviate the effects of waterlogging and salinity.

(d) *Cropping factors*

Another discipline affecting water engineers is imposed by cropping factors referred to in Chapter 2.

Cropping factors probably have a bearing on the vexed question of waterlogging and its attendant evils. Persistent application of irrigation water in excess of that indicated by the time disciplines of potential requirement, as modified by cropping factors, may result in deep percolation and rise of water table. The waste water is not merely lost to the economy, but creates evils which require expensive remedial measures.

The composite effect of these time disciplines, though at first sight mainly of agricultural interest, are of vital importance to the water engineer. In planning storage provisions for supplementary irrigation an understanding of these disciplines is essential.

Sensitivity on run-off

The average long term run-off from the Nile catchment area in Uganda is approximately 6·7% of the rainfall. If, however, as a result of land development or deterioration in the catchment due to bush clearing, or over-grazing for instance, the evapo-transpiration loss should be reduced from 93·3 to 90% the run-off will be increased to 10% of total precipitation. Thus—a reduction of only 3·3% in existing rainfall losses would increase the run-off by no less than 50%. The discharge from Lake Victoria may, therefore, be very susceptible to changes in time due to practices affecting the character of its catchment area.

It is significant that since 1960 the mean yearly level of Lake Victoria, with a surface area of some 72 500 km² (28 000 square miles), has risen beyond anything previously recorded and during the years 1962 to 1964 an average annual rise of approximately 0·4 metres has been recorded, with the highest ever recorded level occurring in May 1964. It is not clear yet to what extent meteorological conditions and to what extent catchment 'deterioration' may have been responsible but the lesson as regards design concepts is very evident, and these would be applicable to most catchments where run-off ratios are relatively low.

It will become increasingly necessary to study the long-term effects on run-off and movement of underground water of soil compaction by heavy farm machinery and the effects of heavier use of chemical fertilizers on the porosity of soils.

The effects of reduction in forest cover and progressive drainage of moorland bogs, acting as sponges, may have to be countered by 'headwater arrestment' measures to avoid increases in flood incidence. (This matter is also referred to in Chapter 9.)

Siltation and depletion of reservoirs

The effects of siltation cannot always be mitigated solely through enlightenment actions by the authorities concerned with the particular catchment. The problems may be inherited from across administrative and political borders and amelioration of silt problems usually involves the co-operation of a number of entities.

The scale of the problem as regards storage depletion in a silt-prone country may be illustrated by reference to Fig. 18 which relates to estimated reservoir conditions in the

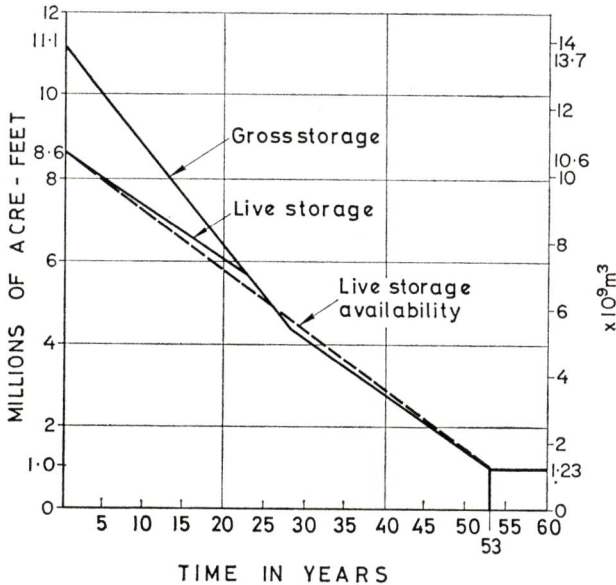

Fig. 18 Estimated storage depletion for reservoirs sited in the middle reaches of the Indus River in West Pakistan.

middle reaches of the River Indus in West Pakistan and shows the expected depletion of live storage capacity of a particular dam over 50 years from $10 \cdot 6 \times 10^9$ m^3 to $1 \cdot 23 \times 10^9$ m^3 ($8 \cdot 6$ to $1 \cdot 0$ million acre-feet). The technical approach to reduce the economic implications of such loss of resources is not always obvious; it would seem that in general the approach must be both protective and productive. The negative, or protective, approach accepts that unproductive structures are required to protect the main, and usually expensive, reservoir (the main breadwinner) as was done in Roman times when small weirs were built across tributaries to achieve regulation for the main stem as may be observed from the ruins on some North African rivers. The principle may be old, but the refinements of application and economic purpose are different. The engineer has to assess the effects in time of the efforts of many other agencies, chiefly agricultural. In addition, he has to mould the designs of these protective but non-productive structures according to foreseeable short term needs. Hence design criteria must be adapted to achieve greatest silt protection at minimum cost. Creative thinking and new design concepts involving not merely the use of new materials but reappraisals of the use of known—even *in situ*—materials is required. Greater calculated risks may have to be accepted, since the consequences of failure of these ancillary minor structures will inevitably have been covered under the various categories of risk provision for the main structure.

The productive approach involves provisions for phased replacement or make-up water for the diminishing main storage. One avenue is to survey in advance the possibilities of valley storage in the tributaries so that clear water may be diverted by pump, tunnel or canal to tributary storage with reduced risk of siltation. Another approach is to so design the structure in the initial stages that it can be raised at a later date, thereby reinstating the storage capacity available. A third approach, with possibly the greatest potential, lies in the utilization of underground storage. It is not commonly realized that in some countries the underground water storage volume exceeds total surface run-off by as much as 100 times.

Increasing use will have to be made of this natural resource for terminal and regulatory storage. In the case of projects for which a pronounced peak is experienced for two or three months a year, it may be costly and difficult to provide maximum canal supply capacity for short periods or infrequent use. Where suitable aquifers are available, it may thus be advantageous, from the aspects of cost and efficient water use, to provide peak supplies by tubewells discharging directly into the distributary system.

Further reference has been made to this subject in Chapter 6.

Groundwater reservoirs are a dynamic resource: receiving and discharging water. Proper management requires adjustment to different time disciplines since there appears to be no direct or simple correlation on the customary annual basis between rainfall and available storage yield. Also, management of the groundwater reservoirs should be carried out with due regard to maintaining the chemical balance in the water.

Retrogression and aggradation

Nowhere in the field of hydraulic engineering are time disciplines so apparent and important as in canal design and operation and in river stabilization. Rivers and canals which have been operated under one set of regimes and have been subject to retrogression suddenly commence aggrading under different regimes of operation or with different inputs of silt type or content. It is remarkable how soon rivers deteriorate, *i.e.* lose their character or flood attenuation potential subsequent to major regime changes.

This phenomenon has been studied on the Indian sub-continent for approximately a century. As yet, there appears to be no sound theory on which firm designs may be based to cover all time disciplines. In the academic field there is still argument as to what precisely causes silt movement and deposit. How much tractive force is involved, how much suction? It may be that the physics of wind-blown sand, as studied in connection with the moving sand dunes of the desert regions, may serve to open up new avenues of research.

In West Pakistan the Marala–Ravi Link canal of some 620 m³/s (22 000 cusec) capacity was originally designed for a slope of 1 : 10 000. Within 15 years it had silted up to such an extent in the head-reaches that discharge capacity was reduced to some 510 m³/s (less than 18 000 cusec); extensive remodelling by dredging and other means was needed to restore the design capacity. Other links designed for 1 : 10 000 have in the same period steepened their slopes to 1 : 7000. The designs for the 620 m³/s Chasma–Jhelum Link between the rivers Indus and Jhelum are based on a design slope of 1 : 10 000 but, to avoid the risk of loss of foundations of bridge and viaduct structures, provision has had to be made for a possible flattening of slope to 1 : 16 000. At the other extreme, to maintain command, provision has had to be made, in control structures along the link, for a steepening to 1 : 8000. This comparatively wide range of design, which cannot be narrowed within present scope of knowledge, is illustrated in Fig. 19. It was felt necessary to make provision at the outlet for initial retrogression of 3 m in the river Jhelum after the Mangla dam located upstream comes into operation and commences to discharge clear water.

Another remarkable illustration of similar time effects is the behaviour of the Colorado River after the construction of Hoover, Davis, Parker, Headgate Rock and Imperial Dams. As clear water was discharged from Hoover Dam into the river which for aeons had adjusted itself to flows carrying large silt loads, profound changes in bed slope took place. Below Hoover Dam the retrogression between 1935 (when the dam closed) and 1947 has been estimated at 95 million cubic metres (124 million cubic yds) involving a bed lowering of some 3 to 4·3 metres. Although Parker Dam, 250 km downstream, has a relatively small storage capacity, the effect on the regimen of the river has been similar, and the retrogression between 1938 and 1947 has been estimated at 115 million m³.

These time disciplines are not confined to situations downstream of the interference. The silt load entering Havasu Lake (Parker Dam) is largely limited to that eroded from the river downstream of Hoover. Available evidence indicates that the river in its natural state has deposited silt in its bed, particularly during low flows. In times of flood, the river overflowed its banks, but since the valley is mainly overgrown by cottonwoods, willows, etc., the suspended silt in the water passing over the banks has been trapped close to the bank, thus forming levees which have tended to confine the river. In the course of time the river and its levees have risen above the adjacent lands.

After the closure of Hoover Dam two distinct developments would appear to have taken place. First, a regular delta was formed in the upper reaches of Havasu Lake. This delta extended perhaps two to three miles upstream from the level pool headwaters of the reservoir. The material forming the delta was then eroded from the river bed below Hoover Dam.

Simultaneously another development took place farther upstream in the middle and upper reaches of the valley. At the time Hoover Dam was closed the river had followed

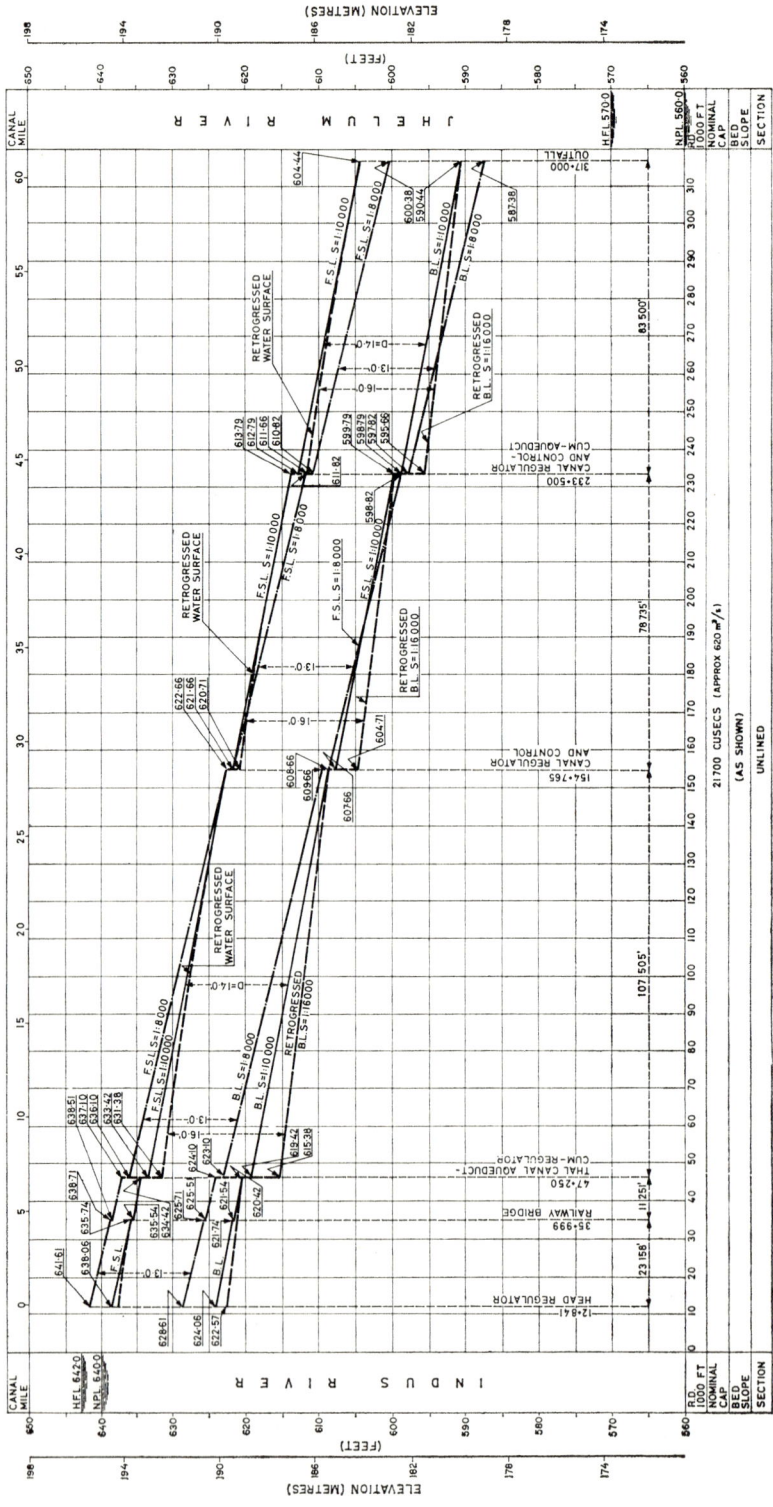

Fig. 19 Chasma–Thelum link canal (West Pakistan) design slope.

essentially the same course for some 50 years or more. The river bed and its natural levees were 3 m to 7·6 m above the adjacent valley lands and the river was ripe for avulsion. With the construction of Hoover Dam, however, the large floods were eliminated and such flows as did occur were not of sufficient magnitude to force the river over its banks on to the lower lands. The silt load of the river, however, had changed, in that not only was the fine colloidal matter no longer present, but the silt itself had become coarser because the material available for erosion below Hoover Dam was coarser than that constituting the river bed before the dam was closed. As a result the upper strata of the river were relatively clear and the silt load was concentrated near the bottom of the stream, possibly in part as bed load.

With the loss of the colloidal material and the absence of silt near the surface of the river, the river lost its ability to construct and maintain natural levees. Whenever an overflow took place it was in the nature of a skimming action of clear water, which precluded the sealing of breaches in the natural levees. An increase of flow in the river, therefore, resulted in the gradual loss of clear water through many small breaks, none of which was of sufficient magnitude to produce an avulsion. The silt, however, which was present in the lower filaments of the stream, remained in the river; but, as the water was lost, the ability of the river to keep this silt load moving was decreased, resulting in accelerated rising of the bed.

The constantly rising water surface elevations in the valley upstream from Havasu Lake have manifested themselves as widespread inundations which in particular have affected the city of Needles and the railway located along one bank of the river. The deterioration of the river had progressed to the point at which the true river bed in 1948 was dry and all water was filtering through the dense vegetation flanking the valley. Large areas which were once dry except at times of high floods became almost impenetrable swamps.

These are short-term effects. In the long term, it has been observed in the Indian sub-continent that accretion upstream of the arterial blockage tends to flatten the slope of the river, but in due course it reverts to its original slope (but with a higher bed level). Similarly retrogression downstream of the obstruction is a passing phase. As the bed level upstream reverts to its original slope more silt is released downstream causing a recovery from retrogression. Generally, the period of recovery is greater than that of retrogression but, as in the case of Khanki on the Chenab, the eventual downstream bed level may be higher than it was originally.

What conclusions of a philosophical nature can be drawn from the manifestations? It appears that nature, whether in the body of homo sapiens, or in the terrestrial body, prefers a balance, and resents interference and will react to any action which disturbs the dynamic balances.

Flood attenuation

Floods are probabilistic in character; on an average, the longer the period of flow observations, the greater is the largest observed flood event. This in itself, and the perils of extrapolation from short-term records, has led to the deterministic approach in design: maximum possible, maximum probable and maximum observed floods. A recent trend has been towards the concept that risks associated with floods with probabilities of the order of 1 : 100 000 should be accepted, since many risks of these probabilities are automatically accepted in everyday life. This is a profound generalization which postulates probabilistic solutions to a probabilistic problem. The degree of risk that can be taken differs not only with the particular type of dam involved, but also with the type of dam and complex of reservoir capacities available or foreseeable both up and downstream of the site under consideration. In the case of major rockfill or earth dams, the cost of spillway protection to be afforded, which must be absolute, is tending to approach half that of the wall itself.

In recent years flood forecasting and routing has become of more and more significance, both from technological and economic considerations. With the rapid expansion in number and size of schemes, the potential damage resulting from failure, or economic loss due to inefficient use of flow, has forced engineers to grapple with the problem of flood forecasting and routing. From entirely empirical methods, whereby designs were based hopefully upon previously experienced maximum flows, the science of flood forecasting has advanced through such methods as statistical projection based upon the past, often insufficient, recorded history of the river, to methods which now embrace the full meteorological and hydrological cycle. The engineer, meteorologist, and hydrologist must now learn to collaborate more and more closely in this important field if society is to benefit to the full from what in the past have been looked upon as somewhat unrelated disciplines.

The next phase is how to design so that the best use is made of the available run-off. Wasteful lowering of reservoir water levels to accommodate floods which may never arise—or risky maintenance of high levels which might lead to overtopping in the event of a large flood—are both to be avoided. This is the point where the engineer must use his ingenuity if many of the benefits of a scheme are not to be squandered; and this is the point where the computer is making a significant contribution.

A great advantage of the electronic computer is that the influence of the various parameters in flood routing can be rapidly assessed and the optimum procedures established. The computer simulation of a river basin enables rapid assessments to be made of design floods, the degrees of control arising from storage at various parts of the river basin and of optimizing reservoir operating techniques. Fed with up to the hour relevant hydro-meteorological data the computer can also be used for flood forecasting thus yielding a running estimate of future flow. The speed with which such information can be obtained is of great value, both as regards economy and security, in the control of complex, highly-developed river basins.

Time disciplines involve not merely assessment of the maximum possible, from which base certain accepted risks in point of time may be evaluated, but also anticipation of the operating procedures required over the life of the project. Designs should be as flexible as possible as regards control of levels and discharges and, on the long-term basis, the aim must be not merely to pass on safely (with reference to the particular structure) the monster flood of rare occurrence, but to endeavour to attenuate these peaks and so to harness to the national economy these wild horses that waste their energy or cause destruction. This may be achieved by provision of single large over-year storage or by successive downstream attenuations of flood peaks in a series of reservoirs. In cases where an extensive reservoir system is not yet foreseen but where suitable ground aquifers are known to be available it is frequently considered to be more beneficial to divert portions of the annual flood peaks for gentle absorption into ground storage than to permit continued discharge to the sea.

With respect to specific design brief reference may be made to the fact that, in overspill conditions, increasing concern is being felt at the possibility that, as in metals, the time discipline imposed by prolonged heavy spilling and the lengthy absorption of energy involved may lead to a form of fatigue in bedrock and/or in the side slope of valleys. This condition is not easy to simulate in model tests, and calls for rigorous observation of prototypes and dissemination of information. Among other considerations, this has led to a careful balancing in design of side and overspill provisions.

Recent studies in connection with armouring of hydraulic surfaces have indicated that shape parameters in the nature of 'packing' factors are involved. Thus a breakthrough may be near with respect to both dissemination and dissipation of energy by armouring of slopes exposed to hydraulic action. It is already known that the carrying capacity in cusecs per foot of surface is greatly affected by the shape of the armouring material and by the manner in which given shapes are exposed to hydraulic action.

Operation

To achieve optimum operation of a given storage facility for multipurpose use it is essential for the operator to be aware continually of the immediate and future demands of each of the purposes being served. It follows that an efficient communications system is a fundamental requirement.

Efficient communications make it possible to use flood storage capacity to its maximum extent to ensure that peaks are attenuated to best advantage and downstream river discharges are reduced to a minimum.

The input in flood control (as in other purpose) reservoirs is highly probabilistic. The aim should, therefore, be to reduce, by design provision and by operating procedures, the probabilistic element in output, partly to minimize damage attributable to water levels and discharges, but also to achieve all-purpose water economy.

Once the shape and peaks of design hydrographs have been established, it is possible to plan hydraulic and operating techniques for flood attenuation in integrated operations. One approach in the past has been to earmark storage capacity purely for flood attenuation purposes. However, such storage space has proved to be increasingly expensive. In some countries good dam sites are scarce, and in most countries the best sites have already been utilized. The economic value of the comparatively rare combination of good dam and reservoir site is illustrated in the case of the Kariba project on the Zambesi river, where the live storage, including flood control, works out at approximately

£0·705 per 1000 m³ (£0·87 per acre-foot), compared with approximately £28·4 per 1000 m³ (£35 per acre-foot) for the Mangla Dam, on the Jhelum river in Pakistan, where topographical and geological conditions are not favourable for large dams.

In these circumstances, the emphasis has shifted towards more refined operating techniques in order to save reservoir space and to use dam structures more economically. Expressed in a different manner, subject to other overriding criteria, economic considerations demand operating techniques which will ensure minimum downstream discharges coupled with minimum rises in lake level during the routing of major floods. The procedures outlined below have been conceived to achieve this ideal.

OVERSPILL AND SPILLWAY GATE CAPACITIES

The routing methods described below are applicable primarily to those dams where flood control is achieved by the combined operation of an ungated overspill and gated sluiceways, the combination of each playing an important part in the routing of floods, and sets out a method of linking spillway gate control with predictions of an incoming flood to achieve maximum flood attenuation through the dam. Since the design philosophies of spilling arrangements differ considerably in earth/rockfill and concrete structures, and because of the space limitations, it is only possible to deal here with operating techniques for concrete dams.

Dams with little or no overspill capacity must, of course, rely on flood storage volume below the dam crest and a flood gate capacity which is a function of the flood storage volume and of the shape and volume of the design flood. Similarly, dams with overspill only or relatively small gated capacity have to be designed for even large flood storage volumes, as the overspill can pass significant flows only after a substantial rise in reservoir level has taken place. For this reason, such dams have to be designed for water levels which may be considerably higher than the full supply level with consequent appreciable extra cost in concrete.

For purposes of illustration, the methods described are applied to a sample river basin having a catchment of approximately 64 700 km² (25 000 square miles) and where the river is liable to flows which vary from about 3 m³/s to a design maximum estimated at nearly 30 000 m³/s (over one million cusec).

The essence of successful flood routing is to transform the basically triangular shape of an incoming flood hydrograph to a trapezium with as long a base as possible. For a given height of dam and practical maximum length of overspill, the optimum spillway gate capacity at full supply level for maximum attenuation through the reservoir can easily be shown to be equal to the overspill capacity at high flood level.

At the arrival of a given flood the gates are opened to give a selected discharge and as the water level rises behind the dam the gates are progressively shut as the contribution from the overspill increases, thus maintaining a constant downstream discharge. Provided the correct gate setting was selected in the first place, the gates will just be fully shut at the moment the reservoir peaks and the overspill discharge is a maximum. Thereafter the level should drop as the peak of the incoming flood will already have arrived some time earlier.

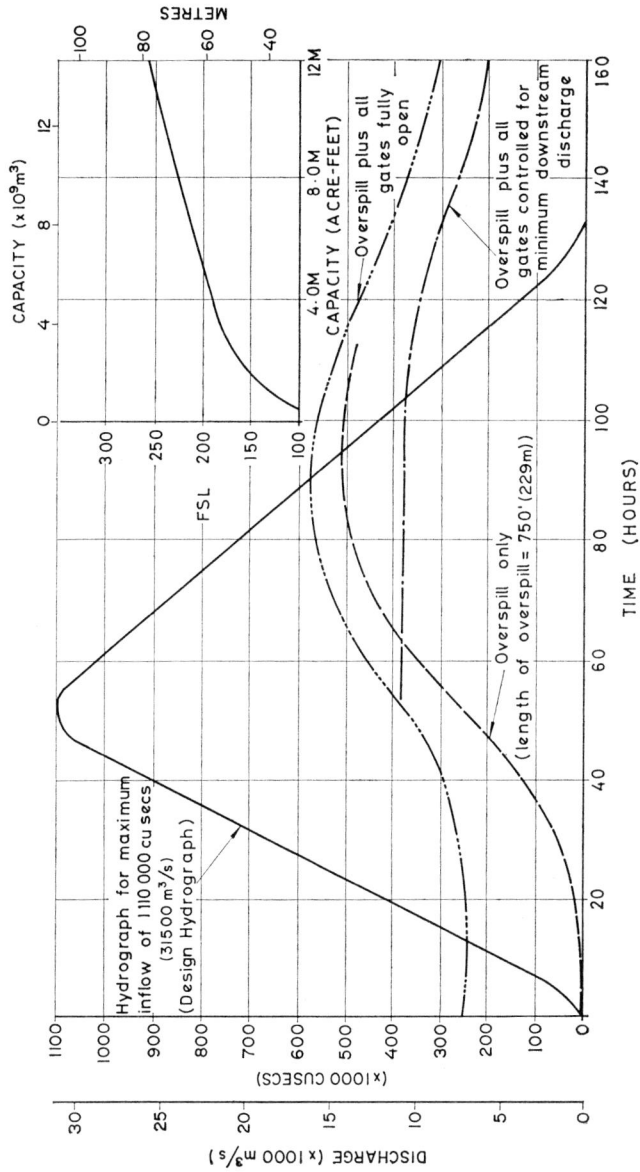

Fig. 20 Flood routing studies for sample reservoir maximum inflow 31 500 m³/s.

The problems facing the gate controller, however, in achieving the theoretical optimum control is difficult as he can have no knowledge of the shape and size of an incoming flood before it has actually arrived, and by then it is too late to adopt the procedure outlined above. A system has, therefore, to be developed which, in spite of the controller's ignorance of the future pattern of an incoming flood, nevertheless enables him to obtain very nearly the theoretical maximum attenuation.

THEORETICAL OPTIMUM CONTROL
In order to measure the degree of success of any practical operating procedure for optimum control, it is first necessary to establish the theoretical control, assuming that at the outset the controller has complete knowledge of the size and shape of the incoming flood.

Figure 20 shows the results of routing a 31 500 m³/s (1 100 000 cusec) flood through the sample reservoir and the attenuation which is achieved by different methods of operation. Figure 21, which shows the optimum discharge out of the reservoir for given incoming

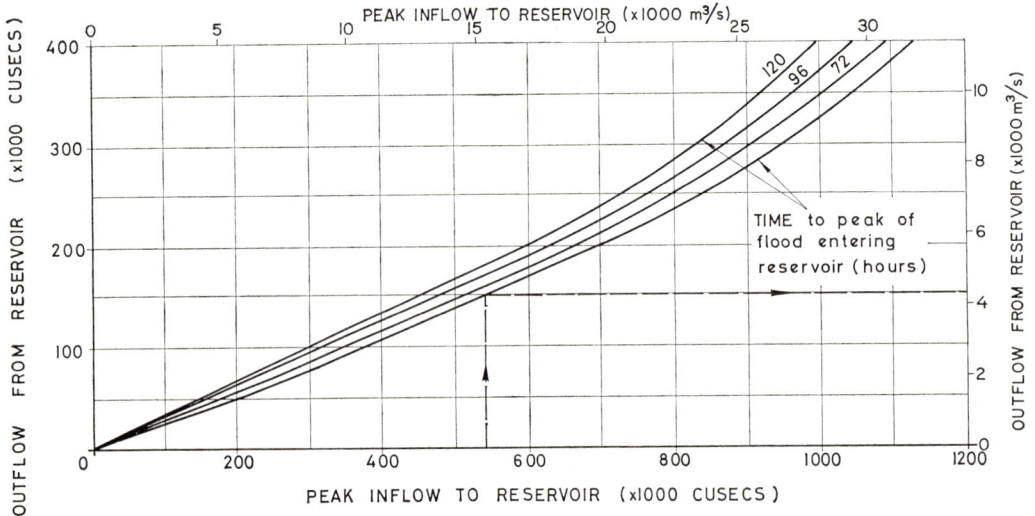

Fig. 21 Optimum discharges from sample reservoir for different peak flows.

floods of various magnitudes and times to peak, has been plotted from similar routings of different magnitude floods. Provided the magnitude and time to peak of the incoming flood is known, Figure 21 can be used directly to select the optimum amount of spilling to adopt.

PRACTICAL OPTIMUM CONTROL WITHOUT UPSTREAM FLOW INFORMATION
Because of disruption of communications or for other reasons, it may be that flood control must be carried out without benefit of any gauging or warning system of im-

pending floods from upstream. In this event, the inflow into the reservoir is determined through the expression

$$q_i = q_o + \frac{ds}{dt}$$

where

q_i = inflow in reservoir
q_o = outflow from reservoir
s = storage in reservoir

The change of inflow into the reservoir is given by

$$\frac{dq_i}{dt} = \frac{d}{dt}\left(q_o + \frac{ds}{dt}\right)$$

$\frac{dq_i}{dt}$ represents the instantaneous slope of the inflow hydrograph at a given time t and this slope is used as one element in the prediction procedure.

The second element used is the 'time to peak' of the unit hydrograph (assumed to be triangular in shape) for the catchment. This time to peak (measured from the commencement of the increase in discharge) is characteristic of a given catchment depending upon its size, shape, slope and rainfall duration, and is assessed on the assumption of uniform rain over the whole catchment. The actual 'time to peak' may, therefore, vary widely from the defined value depending upon the location of the storm on the catchment, but for the larger floods arising from widespread rain the 'time to peak' will tend to approach the defined value.

If T be the characteristic time to peak for the catchment, the peak inflow at time T can be estimated from an observation at any time t through the relationship

$$q_p = q_t + \frac{dq_t}{dt}(T - t)$$

where

q_p = estimated peak
q_t = instantaneous inflow at time t
t = time elapsed since commencement of rise

In practice $\frac{dq_t}{dt}$ is determined by successive determination of the inflow at say 6-hour intervals. If q_{t_1} and q_{t_2} be two successive inflows taken at time $\triangle t$ apart, then

$$\frac{dq_t}{dt} = \frac{q_{t_2} = q_{t_1}}{\triangle t} \quad \text{approximately}$$

The most convenient method of determining q_p will be to keep a running plot of the inflow q_t into the reservoir at intervals $\triangle t$ apart. A straight line extrapolation of the line

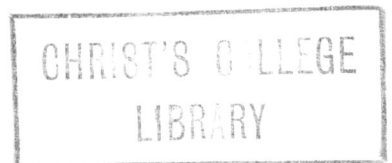

connecting the last two plotted points onwards to the characteristic time to peak T gives the predicted peak of the flood.

The predicted flood peak is then fed into Figure 21, from which the required gate discharge for optimum flood controls is immediately obtained.

At the end of each interval of time $\triangle t$ another prediction of the flood peak is made and the gate discharge can be modified accordingly.

The procedure is continued step by step until the characteristic time to peak T is reached. If the flood peak has not yet arrived, the procedure is continued on the assumption that it will arrive at a time $\triangle t$ ahead. Thus q becomes:

$$q_p = q_t + \frac{\triangle q}{\triangle t} \triangle t = q_t + \triangle q$$

where $\triangle q$ is the increase in flow in the interval $\triangle t$. The revised time to peak is now assumed to be $T + \triangle t$ and the spilling programme, therefore, steps on to the appropriate curve of Figure 21.

For each reservoir a 'threshold' peak predicted inflow should be defined below which little benefit from gate operations can be realized and only when this 'threshold' predicted peak is exceeded should gate operations based on Figure 21 commence. This procedure avoids unnecessary and hasty gate operations for sharp but small total rises in inflow.

An important point in flood routing operations where inflow is calculated from changes in reservoir level, and where the reservoir is of appreciable length, is that the changes in reservoir level must be measured at the null point of oscillation of the lake to ensure seiches do not lead to spurious flood predictions.

A typical flood routing through the sample reservoir based on the procedure outlined above is shown in Figure 22. The 'threshold' value selected in this instance is 2830 m³/s (100 000 cusec).

Cost of money

Whenever consideration of a problem involves money, or its equivalent, and time, interest rates become a vital factor. The principle of charging interest has been known and practised since Babylonian times according to records relating to barter going back to 2000 B.C. A firm of international bankers existed in 575 B.C. with home offices in Babylon which charged a high rate of interest for the use of its money to finance international trade.

Engineers have often been inclined to confuse 'interest' with 'profit'. The former refers to the cost of money in relation to time. Profit may, of course, be reduced to recovering merely the cost of money (interest).

Generally, the interest rate represents to the lender the monetary measure of the sacrifice made by forgoing the availability or use of his capital for one year, and to the borrower the monetary value of the advantage obtained by having the money in advance

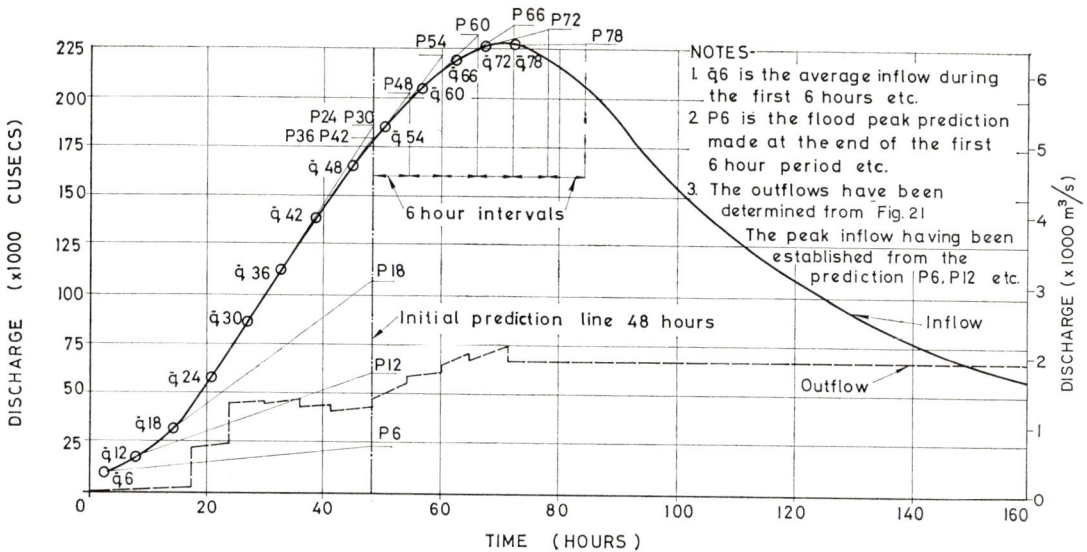

Fig. 22 Typical routing procedure for sample reservoir with no information from upstream.

for the agreed period. The conversion to present values (worths) of future costs and benefits is one of the best means of equating these items in the time disciplines to a common point in time.

At an interest rate of 7% £100 invested today will be worth £107 one year from now. Conversely, £107 to be received a year from now will only be worth £100 today. This is the simplest definition of the principle of present values. For a number of years the question of compound interest arises, involving the standard equation

$$V^n = \frac{1}{(1+i)^n}$$

where

V^n is the present value for any particular year taken from the present,
i is the interest rate.

Standard tables are published from which present value coefficients for one unit of money for any number of years for various interest rates can be read at a glance.

Examples of the use of this technique for calculating discounted cash flows are given in Chapter 4.

A problem in these calculations is to determine or select the 'lives' or periods of effective viability of the various components to a project, *i.e.* the periods of amortization. Civil engineering structures have longer 'lives' than mechanical or electric equipment. Examples from the practice of different countries and authorities are given in Table 14.

Table 14
AMORTIZATION PERIODS ADOPTED BY A NUMBER OF DIFFERENT AUTHORITIES IN RESPECT OF ASSETS OTHER THAN LAND AND WATER RIGHTS

Nature of asset	North of Scotland Hydro-Electric Board	Electricity Supply Board of Eire	Swedish State Power Board	Switzerland	U.S. Bureau of Reclamation and U.S. Corps of Engineers	Government of India
	years	years	years	years	years	years
Tunnels, canals, etc.	80	100	50	50	50	100
Dams, earth, and rockfill	80	100	50	50	50	100
Dams, concrete	80	75	50	50	50	100
Surface power houses	80	75	50	50	50	35
Underground power houses	—	—	50	50	50	—
Steel pipelines	35	20*	50	30	†	40
Gates, steel structures, cranes, etc.—outdoor	35	20*	40	30	†	40
Gates, steel structures, cranes, etc.—indoor	35	20*	45	—	†	40
Turbines, generators, transformers, auxiliary equipment, cables, etc.	35	20*	30	25	†	35

Notes: * These periods were fixed by the Electricity Supply Board at a time when little reliable information as to world experience of the actual lives was available. After the Second World War the Board decided to continue amortization payments on the original basis provided the equipment was still in service. By so doing some provision was made against the likelihood that replacements would have to be made at inflated prices.

† The practice of both the U.S. Bureau of Reclamation and the U.S. Corps of Engineers is to make provision in the estimates of maintenance and operation for the estimated cost of major and minor replacements of those portions of the project which are expected to have a useful physical life of less than 50 years. The method of calculation used has the effect of not taking credit for the residual value of major replacements which would have a remaining useful physical life at the expiration of the 50-year amortization period assumed for the project as a whole.

‡ These rates are laid down in the Indian Electricity Supply Bill (1948) which prescribes that the annual payments into the sinking fund shall be on such a scale that when compounded annually at 3% they shall produce at the end of the period a sum equal to 90% of the original cost of the asset.

OUTLOOK

The items discussed have been selected to highlight, within the space available, a range of issues affecting engineering planning, which are subject to consideration of time

Plate 1 Silt deposits in the Orange River: Republic of South Africa.

Photograph by Struan Robertson, Johannesburg, by courtesy of 'Optima'

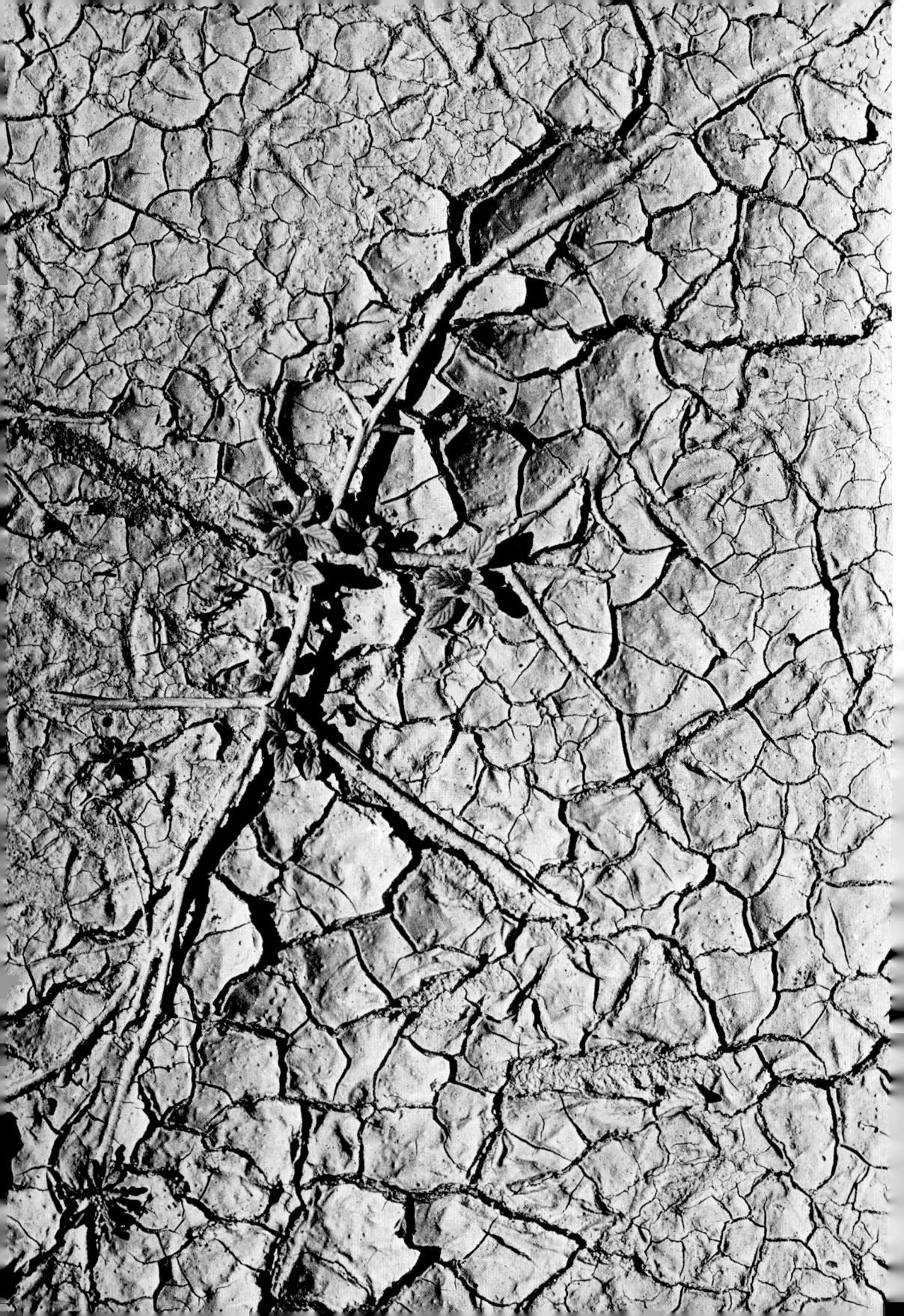

Plate 3 Relief after drought: the thunderstorm.

Photograph by Struan Robertson, Johannesburg, by courtesy of 'Optima'

Plate 4 Reservoir storage capacity almost completely lost through silt: Graaf-Reinet, Republic of South Africa.

controls or effects. The prime objective has been to draw the attention of engineers to the changing patterns and time related problems which require attention in the process of conception and solution of engineering projects.

There are numerous other hydraulic engineering problems which deserve attention. A better understanding of hydraulic roughness is required in order to attack problems associated with energy dissipation, pertinent in both large and very small projects, but very difficult to solve in the latter. More needs to be known about the causes and effects of cavitation in pumps, turbines and on baffle blocks and piers.

It is hoped that these references will serve to encourage reading in a much wider field and to dispel feelings that all is known about the laws of hydraulic engineering. Indeed the contrary is the case—we are but seeing the drop from which to try to visualize the ocean. Some classical theories will be advanced—others will be challenged and modified. There is work for all.

4 Engineering—economic criteria in water resource development

Before a decision can be taken to invest money in the development of an irrigation scheme, some conclusions have to be reached regarding the viability of the scheme and the rate of return on the investment. Feasibility studies are generally carried out under the following broad headings:

1. Technical soundness.
2. Organization and management.
3. Financial arrangements.
4. Economic justification.

In the past engineers have generally been inclined to concern themselves mainly with the first two headings. Recently, however, there has been a distinct change in trend of thought as it became clear that there will not be scope for participation in decision-making processes unless engineers are reasonably familiar with procedures for evaluation of projects for economic and financial viability. This has come about because of the increasing awareness of the impact of irrigation on regional economic growth and fuller appreciation of the fact that, with water as a scarce commodity, irrigation cannot be separated from overall water resources planning, except perhaps in the least developed territories. Thus it is accepted that, in a subject so complex and dynamic, effective planning and implementation of projects can only be achieved through close co-operation between engineering, economics, agriculture, statistics, hydrology, meteorology, education, medicine and other disciplines.

The rate of technological advance, coupled with the dynamic nature of the problems associated with water resource developments, has been responsible for exhaustive reviews of legislation and techniques with a view to effecting improvements in guide-lines and methods of analysis for ranking and selection of projects.

Cost, price and value

It is necessary at the outset to stress the need to avoid confusion in the use of three terms which are not synonymous, but which are frequently used interchangeably when applied to water: 'cost', 'price' and 'value'.

Value of water is the ideal summation of ultimate worth to the community in use and possible re-use of water;
Cost of water is governed by decisions relating to standards of engineering, choice of structures and location, financing and phasing of projects;

Price of water is what the market will bear financially, socially or politically. Being an administrative decision, price may be fixed at any level regardless of cost.

The relationship of cost to value reflects the efficiency of utilization of the resource. However, present procedures for measurement of 'economic value' frequently do not adequately reflect the true value of water in society. For assessment of true value to the community it will be necessary to identify meaningfully such factors as the worth of sustained national nutritional viability and of providing a 'yeast' element for stimulation of regional and national development. These aspects are also referred to in Chapters 1 and 5.

Economic evaluation

Economic evaluation generally entails comparison of 'cost' and 'value', measured from the national and/or regional viewpoint, in terms of the true value of the resources committed to and expected to be derived from the implementation of the project. Thus the main concern is with the total value of benefits to whomsoever they may accrue irrespective of whether they are 'paid for', and not merely with the 'returns' to the sponsoring agency. All transactions which are merely exchanges within the national or regional economy are excluded from cost and benefit considerations.

Financial evaluation

Financial appraisal on the other hand is concerned with the feasibility of recovering the cost of a scheme from the revenues expected to be directly derived from the sale of its services by the organization responsible for its operation, *i.e.* it involves a comparison of 'cost' and 'price'. In most instances this exercise serves as a basis for establishing guidelines for formulating pricing structures, and indicates whether the undertaking will be self-sustaining.

Economic analysis: techniques

The *costs* of a project are the sum of the direct and indirect costs, the former being the assessment of goods and services needed for the implementation, maintenance and operation of the project to make the immediate products of the project available for sale, and the latter representing the costs of further processing and other costs induced by the project.

Similarly, project *benefits* are the sum of the primary project benefits and the attributable secondary benefits. Primary, or direct, benefits are the value of the immediate products or services of the project after deduction of associated costs (all costs other than project costs required for the realization of the benefits). Secondary, or indirect, benefits have been defined as the 'increase in net incomes or other beneficial effects as a result of the project in activities stemming from or induced by the project'.

Economic analysis: investment costs

As the purpose of economic evaluation is to permit comparison between alternative projects or uses of the available resources, it is necessary to include all costs associated with each project, so that the various alternatives may be considered on an equal footing. The cost to the economy of a project is not only that of the major structures involved, but must also include all the other items essential to its successful implementation: cattle breeding centres, roads, processing facilities, agricultural buildings, etc. The fact that such items may not be financed as part of the project does not overcome the necessity for their inclusion in the economic evaluation. Roads, for example, may be essential to transport the produce to market: without such transport facilities the crops would be valueless. The fact that the road is paid for by the Public Works Department, under its general budget, and is not financed by the Irrigation Department from its project budget, does not alter the fact that some part of the country's resources will have to be expended in the construction of the road, in order that the expected benefits from the project may accrue to the economy.

A judgement as to whether a particular cost should be included in the economic evaluation is usually fairly simple in the case of particular items which have to be provided specifically for the purpose of the project. In other cases, particularly where consideration is given to the use of existing facilities or resources, the question may become more problematical. For instance, loss of land tax caused by flooding of land in the reservoir area is a purely financial item, and does not represent a real cost in the sense of a claim on resources. On the other hand, loss of agricultural production in the flooded area is a real loss and should be considered in the project evaluation.

A decision having been taken, as to what items should be included in the investment costs, it is necessary to consider the yardstick that will be used to price those items. Here again, the economic evaluation differs from the financial analysis: in the financial analysis only the actual price paid for the items need be considered. In developing countries, supplementary investments, such as those which are involved in the digging of field channels on the farm, are usually evaluated as costless in economic terms, because such work is usually accomplished during the periods when the farmer would otherwise have been idle: his time and energy are costless because their use in the digging of ditches does not result in a loss to the economy, because there was no other use for his labour. On the other hand, wages paid for skilled labour may sometimes understate its real costs. The wages paid to a farm extension adviser may be considerably less than the real value of his contribution to the economy.

In many countries, the official rates of exchange between the local currency and foreign currencies are only maintained at their official value by means of import restrictions, government regulations, and high tariffs. Thus, the foreign exchange costs of a project may be undervalued in terms of local currency if the official rate of exchange is adopted in the calculations.

While consideration should be given to the real value of the various inputs in the economic analysis, there is little agreement on the extent and ways in which distortion can be evaluated and corrected. Their effort is mainly of use when comparing alternative projects in which the proportion of foreign exchange may be different, or in which the degree of labour-intensive operations vary.

Costs are frequently underestimated in the preparation of project reports or in project evaluation, apparently due in part to bad estimation of project features and in part due to misunderstanding of the concept of economic 'costs'. Apart from an inclination to limit consideration of project costs to those that are expected to be incurred solely by the agency promoting the project there is frequently confusion as to the method of valuing certain project inputs.

The *construction cost* estimate should include the value of all the resources and activities required to design, construct, and put the project into operation, whether these are incurred on the site or elsewhere. In addition to the purely site construction costs these may include items such as: preliminary investigations, design extras, housing for workers, new access roads, airfields and railheads, specialized wagons for transport of bulk cement, increased maintenance costs of existing roads attributable solely to project traffic, new harbour facilities, administrative costs, land acquisition and resettlement costs of the population displaced by inundation of the reservoir areas, additional police, postal, hospital and educational facilities occasioned specifically by and for the undertaking.

The pricing of the project facilities can usually be estimated with some reliability by experienced engineers providing that adequate exploratory work, designs and laboratory tests have been carried out. Such cost estimates are, initially at least, prepared for the purpose of estimating the financial cost of the project, and may require adjustment before they are used for the purpose of economic analysis. The most common adjustments necessary are enumerated below.

In many instances *income tax, import duties and excise taxes* may represent a valuation problem. Provision has to be made for them in the financial estimates because they have to be paid by the contractors building the project. Yet, it is obvious that they do not represent a use of national resources. A reasoned treatment of these items of a purely fiscal nature will vary according to the circumstances. Generally, it is suggested that import duties and excise taxes should be excluded from the economic costs as they are frequently levied either to protect local industry or to control imports, their effect on the government budget being small. Income tax, on the other hand, provides the necessary funds for the majority of all government expenditure in most countries and it is therefore frequently retained as an item in the 'economic' estimate of cost as an offset against the cost of social services provided free of charge by the government to the project, *e.g.* police protection, medical care and public highways.

This approach is undoubtedly very broad in its concept and might be difficult to justify in all cases. In fact, in certain circumstances where unusually large social costs are incurred for the implementation of the project, it may be considered that the income tax allowance is inadequate; in other cases, it may be felt to be more than enough and in such cases further adjustments may be warranted.

Operation and maintenance costs

In the case of irrigation schemes, large costs are likely to be incurred in the maintenance of canals and distributaries. In multipurpose projects these costs are frequently not borne directly by the project agency but by the irrigators themselves or by a separate irrigation authority responsible for distribution of the water provided from the storage reservoir. Because such costs represent a use of resources, even though these may not be reflected in the operational budget of the dam authority, they must be taken into account in assessing the merits of the multipurpose development. In cases where the canal operation and maintenance costs are borne by a separate authority these are usually deducted from irrigation benefits rather than brought into the cost stream of the project itself.

Project benefits

For most multipurpose projects, the measure of the direct benefits from the use of water for agriculture lies in the increase in net value of agricultural products derived therefrom. Even if the direct produce of the project is water, the price of water in the form of water rates, etc., does not reflect its real economic value because the price is frequently established on the basis of essentially arbitrary administrative decisions.

The method of valuing agricultural benefits therefore involves assessment of the gross value of expected agricultural production and of the non-project costs of production to be deducted therefrom to arrive at an estimate of the net value of production at chosen points in time during the project's life. Because the project should only be credited with the incremental benefits attributable entirely to the project, similar estimates of net value of production have to be prepared on the assumption that the project was not undertaken. The direct benefit of the scheme is then equated to the increase in the net value of agricultural output for the 'with project' case when compared with the 'without project' case.

In both 'with' and 'without' cases, forecasts or estimates over the life of the project have to be made of the physical output expected, the values to be applied to this output, and the associated costs of production.

Increases in outputs and yields

The area planted to crops each year is basically determined by water and land availability. Many believe that these are technical questions and can therefore be determined with reasonable accuracy but yet, it is in just these 'simple' technical estimates that great uncertainty may lie. Moreover, experience shows that there is frequently an appreciable lapse of time before full use is made of all project facilities. This is partly because of deficiencies in the necessary supplementary works at the farm level and sometimes because of resistance at the farm level to the introduction of modern methods of farming. Such transitional or development periods are of considerable importance in assessing the rate of return on the project.

The yield per unit of harvested area is subject to a series of influencing factors which include such items as quality of soils and water, use of improved seeds and fertilizer, mechanization, adequacy of pest control measures, drainage and education, availability of agricultural credit and extension services. Because the interaction of such factors is so complicated to analyse and as the problems associated with making reasonable forecasts of the time required to improve or eliminate constraints which may affect any one or more of the items, it has sometimes been found advisable to use two or more projections of the increase in yield in the appraisal of a project. The projections can be made on the basis of optimistic, pessimistic and realistic assumptions regarding availability of all the associated agricultural inputs, and the analysis therefore provides a clear picture of the possible benefits and an indication of the probable results.

Pricing of agricultural output

The value of agricultural output is usually converted to monetary units by using the local market or farm-gate prices, except in the case of export crops for which export prices are considered. The local market prices, however, are frequently distorted by a more or less elaborate set of constraints in addition to which variation in transport costs and malfunctioning of the distribution system can introduce further abnormalities. It is evident that, from a national point of view, investment decisions based on such distorted prices may prove to be misleading. Also, local market prices are subject to seasonal variation unless adequate storage facilities are available to regulate the flow of commodities according to market requirements. Market prices, even if not basically distorted, have therefore to be carefully analysed.

To overcome these difficulties shadow prices, based on import or export prices, are frequently developed whether or not the output in question is intended for export or actually to displace imports. In arriving at the shadow prices or, in the case of export crops, their value, it is necessary to deduct from the export price the cost of processing and transport to the port of shipment. Here again, however, it is necessary to ensure that the real costs are used and not the costs in terms of market prices. For instance, if transport facilities are not fully utilized and the cost of additional transport is consequently very small, only these small costs should be deducted and not the perhaps much higher freight charges actually paid. In other instances the adjustment necessary may be in the reverse direction. The established transport charge may for political or other reasons be such as to favour unduly the conveyance of agricultural commodities. In this case the analysis should take into consideration the *real* costs of transportation and should not be based on the subsidized rates.

However, agricultural *world* markets are also subject to distortion through quotas, subsidies, international agreements, regional preferences, etc., and in the case of commodities of a purely 'national' character relevant international shadow prices may be impossible to establish.

It is therefore accepted that the local market or farm-gate prices will in nearly all cases provide the most meaningful basis of evaluation, and may be adjusted if necessary to take account of any distortions. The degree to which such distortions exist is normally assessed in the light of export or import substitution prices.

It has been found that the most convenient unit of analysis is the farm unit itself. If, as is frequently the case, it is envisaged that there will be several types and sizes of farm within a project, hypothetical farms representative of the various units proposed have to be studied. The basis of study is the farm budget which presents an analysis of the farm's annual balance sheet, *i.e.* an accounting analysis of the income and expenses expected to accrue from and be incurred during a typical year's operation. This farm budget serves first to indicate whether the farming operations will be feasible at all— *i.e.* whether there will be a profit—and, secondly, by the substitution of economic for accounting prices in the budget, to determine the actual value of the national resources that are expected to be employed over a year.

The substitution of economic for accounting prices in such farm budgets is for most items usually simple. For instance, in project analysis the water rates or charges should be eliminated because the water costs are in fact fully reflected in the 'cost' side of the analysis and to keep them in the farm budget would be to double count. Also fertilizer prices may have to be adjusted if the 'accounting' prices used in the farm budget include an element of subsidy. Taxes and imposts are frequently retained as reflecting the cost of providing essential government services which are not detailed or taken account of elsewhere in the 'cost' side of the analysis.

As set out earlier, *direct farm benefits* are the difference between the total value of farm production and the total cost of production expenses including a charge for depreciation and interest on the farm investment, all determined at their *real* value as described in the foregoing paragraphs.

Indirect irrigation benefits, which have also been known as secondary benefits, have been the subject of a great deal of controversy in recent years.

Essentially, indirect benefits are designed to reflect the impact of the project on the rest of the economy. The methods of computing these benefits have been changed a number of times, but they have always rested on certain assumptions and principles.

Frequently they are arrived at by applying a factor to the direct benefit. The factor is dependent on the type of crop and is intended to represent the increase in value of the raw produce as it undergoes a series of manufacturing processes before being marketed.

Thus, crops such as cotton which undergoes a large number of processes before being sold as clothing, usually has a much higher 'factor' than, say, lettuces, which are sold more or less as reaped.

Table 15, reproduced from the U.S. Bureau of Reclamation Manual[42] gives an example of the computation of indirect benefits from farm budgets.

Category A is designed to measure the profits earned on project commodities sold locally. The figure is derived by simply multiplying the total local sales by 5% representing approximately the average profits earned on sales by local distributors.

Category B measures the profit expected to accrue to all *other enterprises between the farmer and the final consumer*. In order to facilitate the analyses of specific projects, the Bureau has derived statistically a set of indirect benefit factors; each factor represents

Table 15

DERIVATION OF INDIRECT IRRIGATION BENEFITS FROM SUMMARY OF FARM BUDGETS FOR ENTIRE PROJECT
(annual values under full development)
(also see Table (10(a) in the Appendix)

Item	With irrigation	Without irrigation	Difference	Factor	Indirect benefit
1. Type of farm	Irrigated farms	Dryfarms & grazing			
2. Number of farms	100	10	(90)		
3. Hectares per farm	65	650			
4. Irrigable hectares	6 500	6 500			
5.	Sales to local wholesale and retail business				
6. Fruit and vegetables	$50 000	—	$50 000	5	$2 500
7. Hay and forage	300 000	$10 000	290 000	5	14 500
8. Sub-total, Benefit A	$350 000	$10 000	$340 000		$17 000
9.	Sales for local and non-local processing, marketing, etc.				
10. Grain	$25 000	$50 000	– $25 000	48	– $12 000
11. Fruit and vegetables	25 000	—	25 000	24	6 000
12. Sugar beets	250 000	—	250 000	26	65 000
13. Seed crops	10 000	—	10 000	10	1 000
14. Dry beans	10 000	—	10 000	23	2 300
15. Soybeans	5 000	—	5 000	30	1 500
16. Livestock (meat)	550 000	25 000	525 000	11	57 750
17. Wool	75 000	5 000	70 000	78	54 600
18. Dairy products	25 000	5 000	20 000	7	1 400
19. Poultry products	25 000	5 000	20 000	6	1 200
20. Sub-total, benefit B	$1 000 000	$90 000	$910 000		$178 750
21.	Purchases for family living and production expenses				
22. Direct farm benefit			$112 500		
23. Less increased perquisites			– 70 000		
24. Increased purchases for family living			$42 500		
25. Increased farm production expenses			840 000		
26. Sub-total, Benefit C			$882 500	18	$158 850
27. Total indirect Benefits, A, B and C					$354 600

Note: Prior to 1967 £1 = $2·81
       After 1967 £1 = $2·40

the ratio of total profits in later processing to the value of the commodity at the farm. By multiplying the list of output quantities by the respective indirect benefit factors the total extra profit is approximated. These two categories of benefits have been called the 'stemming' benefits.

Category C represents the 'profits of all enterprises from supplying goods and services for the increase in farm purchases for family living and production expenses, usually called the induced benefits'. In other words, it is the profit earned on the purchases made by the project. The profit rate has been computed to be an average of 18% for all farm purchases, and this figure is multiplied by the sum of the production expenses and the purchases for family living.

Annual benefit cost ratios

This evaluation technique, which compares the total annual benefits and total annual costs of a project at its *full* stage of development, is no longer considered adequate because it fails to give any recognition to the time distribution of costs and benefits. Thus two projects which are identical except that one reaches full output in one year (such as a power project) while the other takes ten years to achieve full development (such as an irrigation project) would be ranked equally.

Discounted cash flow technique

The most common analytical technique in use today is the discounted cash flow, in which both the cost and benefit streams are discounted at some selected interest rate to arrive at the present worth of the total estimated stream of benefits and costs.

The advantages of the procedure are that not only does it overcome the failings of the annual benefit/cost ratio referred to above, but it also, theoretically at least, takes account of the competition between consumer preferences of the present and demands of the future. The method is simply illustrated by the following formula:

$$V_0 = S_0 + \frac{S_1}{1+i} + \frac{S_2}{(1+i)^2} + \frac{S_3}{(1+i)^3} + \cdots \frac{S_n}{(1+i)^n}$$

in which

V_0 = the present worth of the time stream,

$S_0, S_1, S_2 \ldots S_n$ = the annual amount of cost or benefit in the cash flow,

i = the selected discount rate.

The economic viability of the undertaking may then be expressed either as a ratio (by dividing the total present worth of project benefits by the total present worth of project costs), or in terms of the net present worth by deducting total present worth of costs from the total present worth of benefits. The assumption being that if the benefit/cost ratio is in excess of unity or the net present worth is positive, then there would be a gain to the national economy from the undertaking of the project.

A third method, the internal rate of return method, is merely a present worth procedure whereby a rate of return is calculated for which the present worth of costs equals the present worth of benefits.

While amongst most authorities there is general agreement on the suitability of these techniques, there is considerable debate on their application. The issues which raise the most frequent arguments are:

(a) the interest rate to be used in the calculation;
(b) the period of analysis;
(c) the evaluation of risk; and
(d) what is to be maximized: the ratio of benefits to costs or net present worth of benefits?

The establishment of a correct discount rate is of considerable importance because not only will a change in discount rate alter the benefit/cost ratio or the net present worth of an individual project and thus perhaps determine whether a project is carried out or not but may also affect the relative ranking of projects.

The debate regarding the time span of the economic analysis is very much associated with that of the discount rate. Even though it is generally acknowledged that water resource projects usually have useful life spans in excess of 50 years, the effect on the benefit/cost ratio of including the benefits for years 51 to 100 will be small if the discount rate is high (say 5% or more). On the other hand, if a discount rate of $2\frac{1}{2}\%$ is used the effect of the longer period of analysis may be material as shown in Table 16 which compares the present worth value of the benefits of a typical project, when discounted at two different interest rates.

Table 16
EFFECT ON PRESENT WORTH OF BENEFITS OF CHANGE IN PERIOD ON ANALYSIS AT VARIOUS DISCOUNT RATES

	Present worth of benefits at discount rate of:	
Years	$2\frac{1}{2}\%$	6%
0–50	557·6	235·6
51–100	373·5	22·5
Total:	931·1	258·1
Percentage of total present worth accruing during years:		
0–50	59·8	91·2
51–100	40·2	8·8

The effects of the monetary time disciplines imposed may be summarized briefly as follows:

(a) Increased interest rates make long life alternatives less attractive.
(b) The existence of interest tends to discount the value of future services from long life alternatives.

The main advantages of the D.C.F. methods as a means of assessing whether a project is worth while, or for ranking alternative projects, may be summarized as follows:

(a) It takes account of the effects of the development period on the earnings profile.
(b) It takes into account the effects of phased programming in development.

These advantages are important in irrigation projects with long life structures, and which frequently involve multipurpose projects with different profiles for earning and expenditure streams.

It will be seen that much depends upon the selection of the desirable rate of return. Most industrial firms generally aim to ensure a maximum return on a project not less than they could reasonably expect to earn by investing their money elsewhere. The average return (in real terms) to industrial capital in the U.K. has been estimated by some investigators to fall within a range 6–8% *after tax.*

The expression 'real terms' refers to returns in constant prices, and includes real capital appreciation. The returns in money terms realized in inflationary conditions would normally be higher. However, the engineer will seldom, if ever, be expected to estimate inflationary effects, which even experts find a difficult subject.

For purposes of ranking projects it should not be assumed that the desirable rate of return will automatically be the same for different 'uses'. International Financing Institutions may use one rate for, say, power and another rate for agricultural projects.

Generally the desirable rate of return, or discount rate if the benefit/cost ratio or net present worth methods are being used, will depend on a number of particular financial and economic circumstances prevailing at a particular time in a particular country.

Inputs

In order to derive the most benefit from tools like the Discounted Cash Flow method it is essential, particularly as regards evaluations concerned with irrigation engineering, that the inputs under the different cash streams should be calculated on the basis of valid assumptions. Input calculations are closely linked with land use and cropping intensities. The following definitions and example may help to clarify the basic approach.

The cultivated area can be divided into the *net sown area* and the *non-cultivated area*. The non-cultivated area may have been cultivated in the preceding year (fallow land).

Part of the net sown area frequently carries both a summer and a winter crop. The *total cropped area* is therefore larger than the net sown area.

$$\text{Land use intensity} = \frac{\text{net sown area} \times 100}{\text{culturable area}}$$

$$\text{Cropping intensity} = \frac{\text{total cropped area} \times 100}{\text{net sown area}}$$

Table 17 illustrates typical cropping intensity data from a recent survey undertaken in West Pakistan.

Table 17
WEST PAKISTAN: CENSUS LAND USE AND CROPPING INTENSITIES[39]
(also see Table 11(a) in the Appendix)

| | | Farm size | | | |
		Under 2 hectares	2–9·9 hectares	10 hectares and over	Total
Culturable area		173	1025	676	1874
Cultivated area	× 1000	161	945	593	1699
Net sown area	hectares	139	827	504	1470
Total cropped area		169	995	589	1753
Land use intensity		80	81	75	78
Cropping intensity		122	120	117	119

Typical farm accounts in the Punjab region of West Pakistan are illustrated in Table 18 based on census figures for the year 1954–55 covering Owner Farmers and Tenant Cultivators.

It will be seen from Table 18 that water charges (not to be confused with the *value* of water) represent relatively small proportions of the accounts.

A typical example of the discounted value of net agricultural benefits and of 1000 m³ of stored water, for assumed low, medium and high input levels for various discount rates, is provided in Table 19.

Estimates of net livestock output per cropped hectare are not easy to obtain and vary considerably from place to place. A typical example is given in Table 20.

In assessing the true economic benefits of a project, it is essential to check for nutritive viability. There is little point in growing lucrative cash crops for export unless there is a favourable balance on foreign exchange for essential foodstuffs which may have to be imported. A typical survey of projected requirements is illustrated in Table 21.

Financial analysis—techniques

The objective of financial analysis is to determine the extent to which project costs are likely to be recovered from sale of the services to be provided by the project. Initially, however, the procedures involved in financial analysis serve to establish the scale of charges that should, or may, be levied for the services provided. They also serve to clarify

Table 18
FARM ACCOUNTS IN THE PUNJAB (WEST PAKISTAN)[39]
(also see Table 12(a) in the Appendix)

Item	Owner farmer 1954–55 U.S. dollars	Tenant cultivator 1954–55 U.S. dollars
Number of farms	8	9
Area in hectares	58·2	48·9
Cultivated area in hectares	50·8	47·7
Cropped area in hectares	60·6	54·3
Permanent labour family members	15	16
Hired	5	2
Number of pairs of bullocks	14·5	11
Gross income per cultivated hectare in U.S. dollars	121	93·5
Actual expenditure per cultivated hectare in U.S. dollars:		
Manual labour—Permanent	8·76	1·10
Casual	5·98	2·94
Bullock labour	16·38	12·82
Seed	4·35	4·53
Implements	1·67	1·44
Artisans	1·56	0·84
Manure	0·62	—
Rent	—	32·50
Land revenue taxes and water charges	4·72	1·62
	44·04	57·79
Net income per cultivated hectare in U.S. dollars	82·00	35·40

Table 19
DISCOUNTED VALUE OF NET AGRICULTURAL BENEFITS
AND UNIT VALUES OF STORED WATERS[39]
(also see Table 13(a) in the Appendix)

Discount rate %	Input levels	Benefits U.S. dollars $\times 10^9$	Value of stored water dollars $\times 1000 \, m^3$
4	Low	0·925	5·32
	Moderate	1·218	6·97
	High	1·617	9·32
6	Low	0·546	3·11
	Moderate	0·715	4·12
	High	0·945	5·44
8	Low	0·336	1·94
	Moderate	0·441	2·51
	High	0·588	3·34

Table 20
WEST PAKISTAN: NET OUTPUT OF LIVESTOCK PER CROPPED HECTARE
(also see Table 14(a) in the Appendix)

Gross output per cropped hectare	U.S. dollar	33·75
Cost per cropped hectare:		
Feed	$11·41	
Concentrates and salt	2·60	
Hired labour	1·04	
Veterinary	1·29	
Miscellaneous	0·52	
Interest and depreciation	5·19	
	Total: $22·05	
Net livestock output per cropped hectare	$11·70	

Table 21
WEST PAKISTAN: FOOD REQUIREMENTS FOR A 2500 CALORIE/DAY DIET
(also see Table 15(a) in the Appendix)

Description	1965	2005
	in millions	
Population:	51·90	125
Adult male units:	41·50	100

	Standard consumption per adult per year in metric tonnes	Food requirements in thousand metric tonnes	
Cereals	0·1686	6990	16850
Pulses	0·0317	1312	3165
Sugar and gur	0·0211	874	2107
Vegetables	0·0632	2820	6320
Fruits	0·0317	1312	3165
Fat and oils	0·0158	656	1578
Milk and milk products	0·0843	3495	8420
Fish, meat and eggs	0·0263	1093	2635

the extent to which, in the case of a multipurpose project, costs are incurred in the provision of project features for which no financial return may be expected, and the proportion of the total investment associated with each of the particular purposes of the project.

It is in this latter field of 'cost allocation' that most difficulties arise as a result of the multiplicity of conceptual problems inherent in an approach to cost allocation, and the marked effect it may have on the establishment of tariffs and charges and the degree to which local or regional financial participation in the project may be dependent on it.

Allocation of costs

There are several different approaches possible to the allocation of costs of a multipurpose structure:

—alternative justifiable expenditure
—priority of use
—benefits
—use of facilities
—equal apportionment and
—vendability

The alternative-justifiable-expenditure approach would require that the cost of the alternate single-purpose project or the maximum justifiable expenditure, whichever is less, be taken as the relative value of the cost assignable to a single-purpose. Application of this method requires that sufficient data be available to estimate the cost of substitute single-purpose facilities with reasonable accuracy. It also assumes that a multipurpose project can provide services more cheaply than individual single-purpose projects.

The priority of use approach recognizes that one or more functions of a scheme may have a priority in its claim for space in the reservoir. For example, priority may be given to one use, such as irrigation, and releases for other uses, such as power and navigation, may have second or third priorities until the requirements of the first priority are fully met. Therefore, after the specific costs of the facilities required for carrying out the project purposes assigned to each project function have been determined, the residual costs are then allocated according to priority of use.

The benefit criteria approach would require that the total cost of a multipurpose project be divided in proportion to the benefits expected to accrue from each purpose.

The use-of-facilities approach would allocate the costs on the basis of amount of use of the facility for each purpose of the scheme. This method assigns a share of the cost of a joint facility, such as a dam, proportionate to the respective uses of the joint facility.

The equal-apportionment method requires that costs of the joint facilities be allocated equally among the various uses to be served by the project.

The vendability method is based on the theory that under conditions of perfect competition, a producer of joint products would so arrange output that long-term revenues from joint products should at least equal the fixed costs plus the variable cost of the combined output. Thus, the cost of joint facilities is divided by first subtracting from revenues obtained from each project feature, the specific cost incurred for that feature and secondly, by capitalizing the remaining revenues at an appropriate rate of interest. The result obtained from the second step should equal or set a ceiling on the amount of the cost of the joint facilities to be assigned to a particular feature.

Some of the above concepts are incorporated into the method most commonly adopted today for cost allocation: the Separable Costs–Remaining Benefits Method, proposed by the sub-committee on Evaluation Standards of the Inter-Agency Committee on Water

Table 22

COST ALLOCATION SUMMARY OF THE ANIMAS–LA PLATA PROJECT: SEPARABLE
COSTS–REMAINING BENEFITS METHOD[43]

(100 years—$3\frac{1}{8}$ percentage interest. Unit: $1000)

Item	Irrigation	Municipal and industrial	Recreation fish and wildlife enhancement	Total
Construction costs	—	—	—	$109 493
Interest during construction $3\frac{1}{8}$ percentage	—	—	—	7 190
Present worth of 100 years' annual operation, maintenance, and replacement costs	—	—	—	9 768
Costs to be allocated (present worth)	—	—	—	126 451
Benefits:				
Annual value	$5 541	$1 840	$184	7 565
Present worth	169 139	56 166	5 617	230 922
Alternative single-purpose cost (present worth)	106 273	41 627	*	—
Construction (present worth)	92 370	25 150	—	—
Interest during construction (present worth)	6 150	1 520	—	—
Operation, maintenance and replacement:				
Annual value	254	490	—	—
Present worth	7 753	14 957	—	—
Justifiable expenditure	106 273	41 627	5 617	—
Separable cost	73 182	17 540	2 720	93 442
Construction	62 540	15 890	1 243	79 673
Interest during construction	4 110	1 040	42	5 192
Operation, maintenance and replacement:				
Present worth	6 532	610	1 435	8 577
Annual value	214	20	47	281
Remaining justifiable expenditure in excess of separable costs	33 091	24 087	2 859	60 037
Remaining joint costs (percentage of total)	55·1	40·1	4·8	100·0
Allocated joint costs:				
Construction	16 431	11 958	1 431	29 820
Interest during construction	1 101	801	96	1 998
Operation, maintenance and replacement:				
Present worth	656	478	57	1 191
Total allocation:				
Construction	78 971	27 848	2 674	109 493
Interest during construction	5 211	1 841	138	7 190
Operation, maintenance and replacement:				
Present worth	7 188	1 088	1 492	9 768
Annual value	235	36	49	320

Note: * Higher than benefits

D

Resources in the U.S.A. Under this method each purpose is first charged with its separable costs which is the difference in the total project cost resulting from the inclusion or deletion of any purpose. The difference between the *sum* of the separable purpose costs and the *total* project cost, constitutes the common costs chargeable to all purposes served. This residual is distributed among purposes in accordance with the excess of its benefits over its separable costs. Alternative costs take the place of benefits when they are less.

A typical example of the application of the Separable Costs–Remaining Benefits Method is given is Table 22.

Repayment schedule

The end result of the financial analysis is to tabulate year by year the expected revenues to be derived from the sale of the project services at the established prices, and to compare these with a similar cash flow of the costs of constructing and operating and maintaining the project.

For the purposes of this cash flow the projections of demand for project services established during the course of the economic analysis are used.

The treatment of interest during construction requires special consideration. Interest at the borrowing rate is usually added to the project cost to arrive at the total capital sum invested by the time the works are completed. However, in some cases initial revenues are inadequate to cover recurrent expenditures, including loan servicing and hence the investment must be increased by the cumulative deficit.

Intangible benefits

Placing a value on certain benefits in multipurpose developments presents problems in that there is not available as yet a significant unit of measurement. This applies for instance to recreational benefits and to the impact of water on regional development. Subject to further research it is suggested that, in an addendum to the benefit/cost or D.C.F. calculations, such *intangible* benefits should be explicitly enumerated and described, with particular reference to probable effects on selected economic growth indicators.

General conclusions

The general trend of thought as regards the use of economic and financial analyses may be summarized as follows:

(a) The basic justification of a scheme or plan should rest upon the economic evaluation supplemented by judgement of other factors such as regional impact, employment opportunities, national security including nutritional viability. Thus a project would be considered for selection or ranking if the present worth of the national resources committed to it are less than the present worth of the national benefits expected to be derived from the project.

(b) The role of the financial analysis would appear to be to serve as a basis for establishing guide-lines for pricing structures. This would cover such items as charges for irrigation water which, on a global basis, would appear to be inadequate to meet O. & M. costs, and the prices for the sale of the project commodities. Such analyses would make it possible to clarify explicitly the extent of financial subsidies required, if any, to achieve specific overall objectives.

It is stressed that the complex issues of economic criteria for engineering projects cannot be covered in a single chapter. The foregoing is intended only to highlight briefly some pertinent aspects in order to indicate the importance of economic and financial considerations in the decision-making processes and to suggest a wider field of reading on this subject. To this end there is listed, after the Bibliography, a supplementary list of publications covering various aspects of this subject, and a simple example of a present worth/discounted cash flow calculation for a water resources project is set out below.

Costs (R. × 10³)

| Year | Investment | | Operation & Maintenance | | Total annual costs | Present worth at discount rate | | |
	Dam and power plant	Irrigation	Dam and power plant	Irrigation		of 6%	of 10%	of 11%
1	1070	79	—	—	1149	1083	1044	1034
2	5000	211	—	—	5211	4632	4304	4226
3	5700	215	—	—	5915	4962	4442	4324
4	4300	261	20	20	4641	3676	3169	3054
5	1210	184	80	40	1514	1131	938	898
6	—	—	500	60	560	394	316	299
7	—	—	500	60	560	372	287	269
8	—	79	300	60	439	275	204	190
9	—	211	200	60	471	278	199	184
10	—	215	120	60	395	220	152	139
11	1420	261	120	80	1881	989	658	596
12	2800	184	120	100	3204	1589	1018	913
13	—	—	150	120	270	126	78	69
14	—	—	150	120	270	119	71	62
15	—	60	150	120	330	138	79	69
16	—	200	150	120	470	185	102	88
17	—	240	150	120	510	189	100	86
18	—	280	150	140	570	199	102	86
19	360	200	150	160	870	287	142	119
20	1070	60	150	180	1460	454	216	181
21	—	200	170	180	550	162	74	61
22	—	240	170	180	590	163	72	59
23	—	280	170	200	650	170	72	58
24	716	200	170	220	1306	321	132	106
25	2140	—	170	240	2550	591	235	186
26	—	—	190	240	430	94	36	28
27	—	—	190	240	430	89	33	25
28	360	—	190	240	790	154	55	42
29	1070	—	190	240	1500	276	94	72
30	—	—	200	240	440	76	25	19
31	—	—	200	240	440	72	23	17
32	—	—	200	240	440	68	21	15
33	—	—				64	19	14
34	—	—				60	17	12
35	—	—				57	16	11
36	—	—				54	14	10
37	—	—				51	13	9
38	—	—				48	12	8
39	—	—				45	11	7
40	—	—	↓	↓	↓	43	10	7
14–50	—					319	60	40
				Total		24275	18665	17692

Power (thermal alternative)			Irrigation			Present worth at discount rate		
Invest-ment	O. & M.	Fuel	Net increase in N.P.V.	Flood control	Total annual benefits	of 6%	of 10%	of 11%
—	—	—	—	—	—	—	—	—
200	—	—	—	—	200	178	165	162
700	—	—	—	—	700	587	526	511
1 210	71	—	71	—	1 352	1 071	923	889
2 200	159	143	168	246	2 916	2 178	1 808	1 729
800	159	163	285		1 653	1 164	932	883
1 200	159	193	308		2 106	1 400	1 080	1 013
—	203	225	329		1 003	629	467	434
—	203	259	345		1 053	622	446	411
200	203	296	359		1 304	728	502	459
800	203	336	500		2 085	1 097	729	661
1 410	203	330	648		2 637	1 308	838	752
—	242	374	805		1 667	780	762	428
—	242	413	822		1 723	761	453	398
—	242	475	840		1 803	752	431	377
—	242	530	858		1 876	737	407	353
200	242	567	876		2 131	791	420	360
800	242	630	1 043		2 961	1 036	530	450
1 200	242	650	1 216		3 554	1 173	579	487
—	290	707	1 395		1 638	509	242	203
—	290	770	1 422		2 728	802	368	303
200	290	876	1 438		3 050	845	372	305
800	290	948	1 616		3 900	1 018	433	353
2 400	290	1 020	1 797		5 753	1 415	581	470
3 000	325	1 110	1 882		6 563	1 523	605	483
4 210	325	1 130	2 004		7 915	1 733	664	525
—	484	1 260	2 023		4 013	831	306	239
—	484	1 370	2 042		4 142	808	287	223
200	484	1 485	2 061		4 476	823	282	217
900	484	1 610	2 081		5 321	926	305	232
3 000	484	1 910	2 100		7 740	1 269	402	304
—	512		2 117		4 785	737	226	169
—			2 134		4 802	701	206	153
—			2 148		4 816	659	188	138
—			2 162		4 830	628	171	125
—			2 177		4 845	591	156	113
—			2 191		4 859	559	143	102
—			2 206		4 874	528	130	92
—			2 220		4 888	503	119	83
—			2 234		4 902	476	108	75
—			2 304		4 972	3 683	681	453
				Total		38 559	18 973	16 117

5 The impact of irrigation on regional economic growth

Distinct trends of thought in water resources engineering are emerging as statesmen or policy-makers become increasingly aware, firstly, of the potential, through irrigated agriculture, for generating economic activity at regional and national levels, and secondly, of the consequential opportunity, through such economic activity, for promoting peaceful inter-regional relationships. One example of the extent of such global interest was the 'Water for Peace' conference held in Washington, D.C., during May 1967, attended at ministerial, expert and observer levels by representatives from some 75 countries resulting in the presentation of approximately 700 papers on this subject.

It has become clear that, except in the least developed territories, irrigation cannot be separated from overall water resources planning. Economic (as distinct from purely financial) evaluation of storage, distribution and infrastructure projects in relation to resultant benefits automatically involves public funds and, therefore, national considerations, having regard to such factors as tax structures, economic defence aspects and the criterion that the greatest benefit must accrue to the greatest number.

Since World War II, economists have entered the field and have rendered great service in attempts to devise methodologies and criteria for evaluating alternative uses of water, as well as determining, usually by a continuation of statistical and theoretical considerations, the value of water for a particular purpose in a particular area for given crop patterns at a particular point of time.

At first there was considerable opposition by engineers to the participation by economists in exercises of conceptual engineering. While there is still some residual resentment, there is a growing school of advanced engineers who recognize that the subject covers a wide spectrum of interests and that, while engineering is the corner-stone as regards planning and implementation, economics, agriculture, statistics, hydrology, meteorology, education, medicine and numerous other disciplines are involved. Moreover, a growing awareness of the dynamics of the emergent problems, in relation to available technical and financial capacity for providing solutions to these problems, has served to focus attention on the vital need for co-operation right from the early planning stages of water resources projects between engineers and all the other disciplines involved.

Much, however, remains to be done as regards refinement of methodology and criteria for evaluation of economic benefits. There is, for instance, at present still no reliable means for measuring the effects of institutional constraints, established rights, nutritive factors and catalysts for generating wealth.

The following illustrates briefly, with data from countries with widely different conditions, two major factors which influence the assessment of *value* of water to the economy:

(a) National or regional status of nutritional viability;
(b) The *impact* or *yeast* factor of irrigated agriculture in stimulating economic growth within the project area and for promoting outward radiating influences on adjacent communities.

NUTRITIVE FACTORS

Whereas on the one hand the world is bulging from accelerating increases in population, on the other it is shrinking as regards all forms of communication. In combination, these twin dynamic trends generate pressures to equalize standards of living and render it increasingly difficult to isolate hunger geographically, with subsequent great potential danger for all mankind.

Geographically, at present, food problems centre primarily in the Far East and in Communist Asia where the dense and growing populations increasingly jeopardize the food supply. Thus the Free Far East had two-thirds of the wheat shortage projected for 1962–66 and nearly one-half of the animal and pulse protein and fat deficit. This region represents 42% of the population of the regions with diet deficit and has 60% of the deficit. Communist Asia has most of the rest of the animal protein and fat shortage and, in spite of large purchases of grain, still has a considerable calorie gap (Fig. 23).

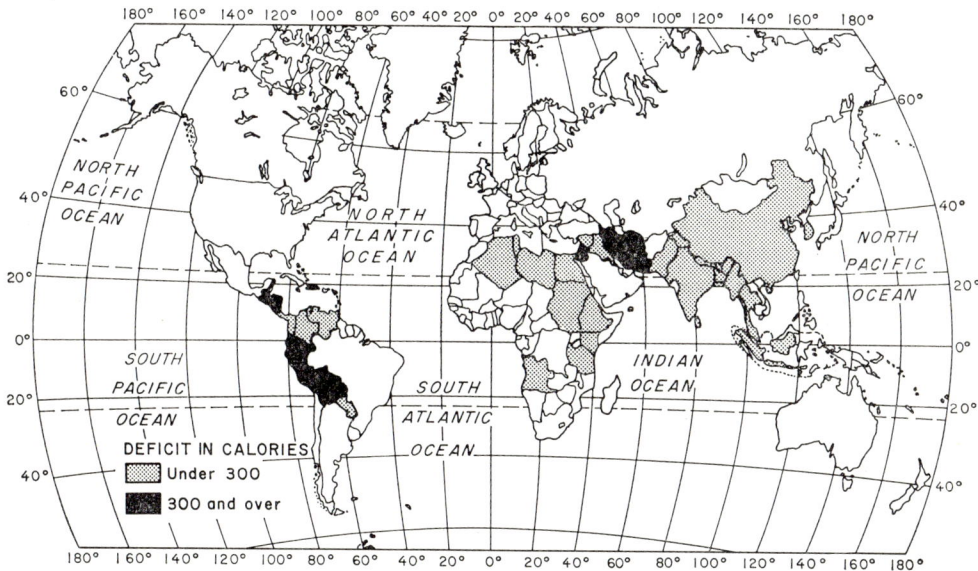

Fig. 23 Calorie deficiency in *per capita* daily average diet: 1958.

Unfortunately, the dynamic nature of the problem is such that there is no room for complacency. Figure 24, based on recent World Bank surveys, illustrates that the

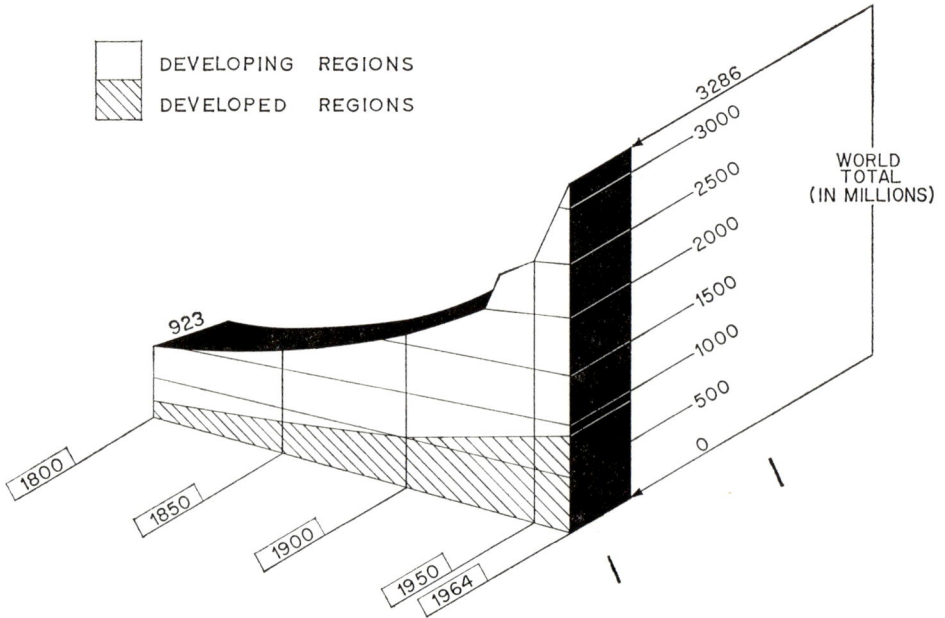

Fig. 24 Present and projected world population.

projected population of the Far East, including mainland China, will, by the year 2000, exceed by far the combined total of the rest of the world. What is most significant is the acceleration in population pressure in the less advanced regions. Between the years 1800 and 1964 the world population rose from 923 million to 3286 million, of which only 900 million are located in regions of advanced development.

Diet deficiency has pronounced impact on national economy as regards output per man hour, expectancy or life, health requirements, import of foodstuffs, hence foreign currency problems and, therefore, political alignments.

As far as the capacity for work is concerned, the main results of a poor diet are that the body avoids effort, there is a lowering of resistance to disease, and the accident rate rises. The symptoms of the avoidance of effort are lethargy and a lack of initiative and drive.

The F.A.O. have decided upon a Reference Man, who is 25 years old, healthy, weighs 65 kg, and lives in a temperate zone with a mean annual temperature of 10 °C. For the Reference Man, the following calorie requirements have been drawn up.

However, just as the requirements for the Reference Man are not generally applicable, so terms like sedentary are of restricted usefulness. One occupation, such as farming, can include jobs which involve widely varying degrees of activity. Also, conditions vary widely from country to country; farming in a highly mechanized area is very different from farming in relatively primitive conditions.

Table 23
CALORIE REQUIREMENTS OF F.A.O.
REFERENCE MAN

Degree of activity	Calories/day
Sedentary	2800
Moderate	3200
Heavy	4400

The benefits of supplementing a poor diet have been demonstrated in a recent experiment with a group of Ruhr coal miners:

Table 24
EFFECT OF CALORIE INTAKE ON
WORK OUTPUT
(also see Table 16(a) in the Appendix)

Calories available for work	Output (metric tonnes)
1200	6·81
1600	9·55
2000	10·98

The effects on productivity of a nation arising from increased expectancy of life, high standards of nutrition and hence greater overall national efficiency, may be illustrated by an over-simplified example comparing a small country enjoying high standards of nutrition and health with a large country suffering from cumulative effects of chronic diet deficiency (Table 25).

Table 25
EFFECT OF NUTRITION ON NATIONAL LABOUR FORCE

Country	Population (million)	Expectancy of life (years)	Effective working life (years)	Overall efficiency %	Effective man years (million)
A	400	35	20	25	2000
B	50	75	60	75	2250

Although not a meaningful comparison in every sense, it is interesting to note from Fig. 25 the trend: high G.N.P. (Gross National Product) *per capita* for the food adequate regions of North America and Europe compared with the relatively low G.N.P. *per capita* for countries like India and Pakistan.

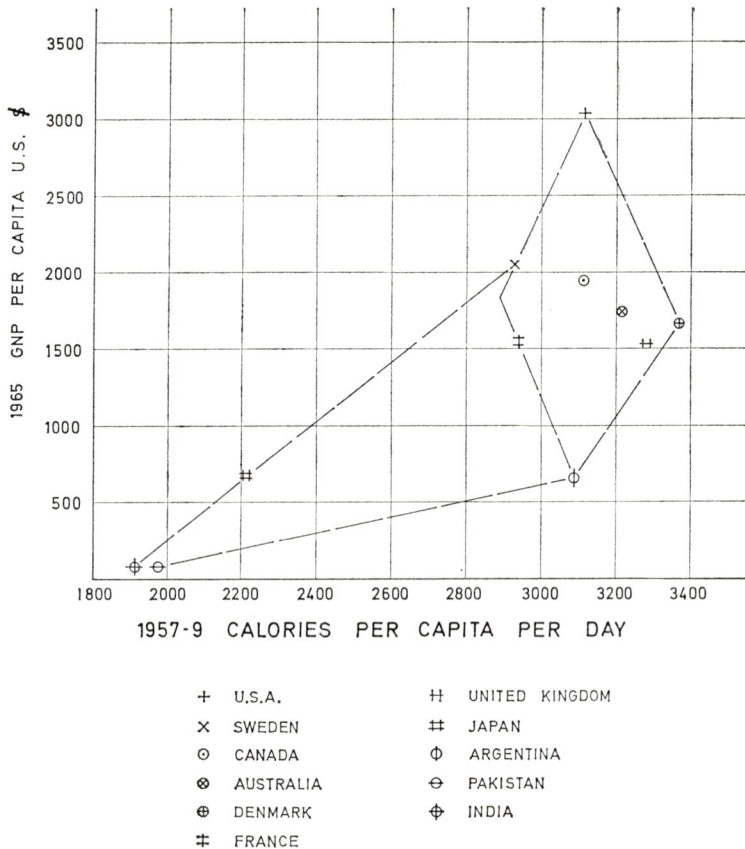

Fig. 25 Relationships of gross national product and national average calories *per capita* per day.

Whether considered from the food or the cash crop angle, problems associated with irrigated agriculture are dynamic in character. The term developing country denotes a dynamic process. No country is ever developed. A general indication of the movement up the status graph may perhaps be given by comparing income distributions within the population at various earning levels. Figure 26 shows such a comparison for Calcutta, France and the U.S.A. The vector trend in development is generally from an agricultural towards an industrial economy.

Most developing countries depend on agriculture which constitutes approximately 60% of their gross national product and provides the basis of living for about 80% or more of their people. As the country develops the agricultural sector provides initially the raw materials for industrial growth, the means for mobilizing capital and the facilities for earning foreign exchange.

However, it is questionable, for reasons already mentioned, whether the measurement of benefits only in monetary units provides a fair representation of the value of water on

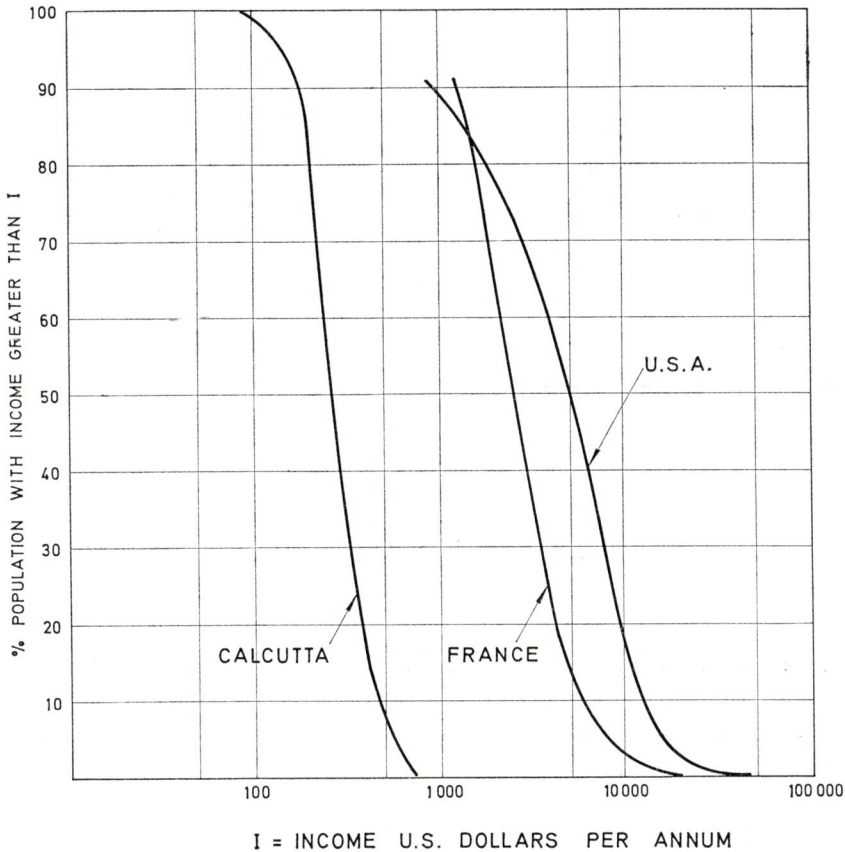

Fig. 26 Comparison for three countries of the distribution of incomes within the population.

both a short- and a long-term basis. The economic efficiency of the community clearly depends on diet standards and hence there is, for each environment, a critical nutrient level, below which the prime motive of the agrarian society must be preservation, and only above which it can be fully profit motivated. This consideration is of prime importance in forward planning.

Added impetus is given to such considerations by the alarming trends revealed in recent surveys by the U.S. Department of Agriculture. The results are embodied in the two sets of graphs in Fig. 27. It is seen that whereas the total and *per capita* food production for the more developed countries show a steady increase from 1957–59 (= 100%), the *per capita* food production for the less developed countries listed over the same period increased very tardily and, since 1963, there has been an alarming drop back to 1954 levels. The causes are to be found in the rate of population increase and bad agricultural practices.

1957-59 = 100

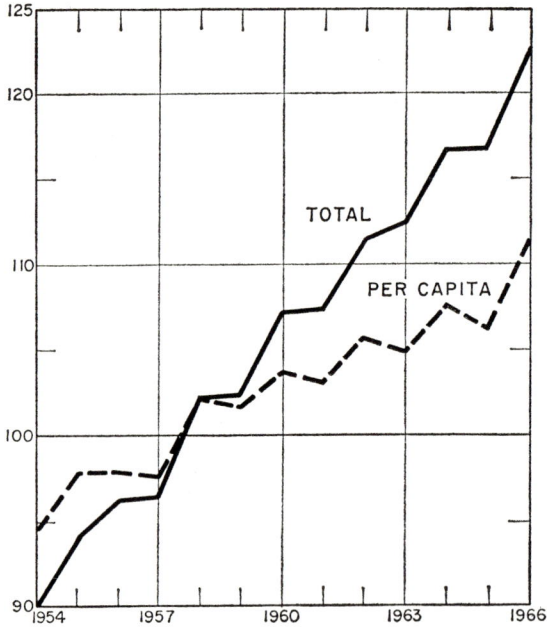

DEVELOPED COUNTRIES

United States, Canada, Europe, U.S.S.R.,
Japan, Republic of South Africa,
Australia and New Zealand

1957-59 = 100

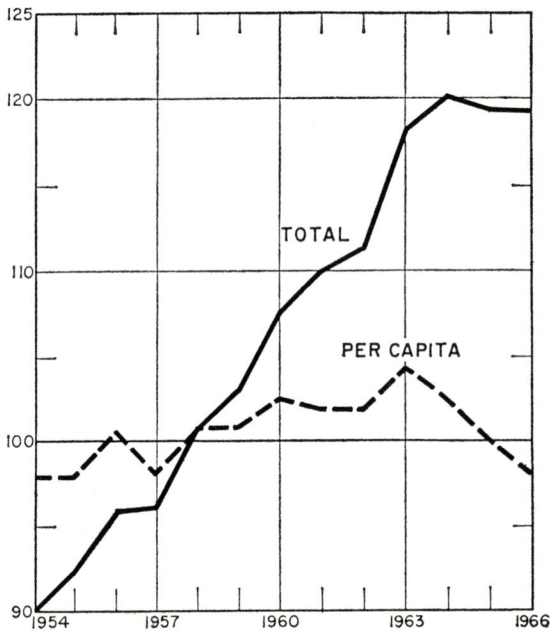

LESS DEVELOPED COUNTRIES

Latin America, Asia except Japan and
Communist Asia, Africa except Republic
of South Africa

Fig. 27 World food production 1954 to 1966.

Examples: West Pakistan

The most modern illustration of the impact of irrigation on a regional economy and the relevance of the foregoing lines of thought, applied to a developing country with predominantly rural economy and diet standards still below critical nutrient level, is to be found in West Pakistan.

As mentioned earlier, the implementation of the vast Indus Basin project assured the basic replacement of waters in accordance with the sharing arrangements enshrined in the treaty between India and Pakistan. In addition to elements of development, such as storage and hydro-electric capacity, it provided the springboard for future development of water resources. With this in mind, there was mounted in the early stages of the Indus Basin project a massive study, probably the first of its type, carried out by a group of World Bank staff assisted by international Consulting Engineers. A few examples will serve to illustrate some of the valuable lessons which emerged from this study.

Nutritional Viability
On the basis of a national average critical nutrient level of 2200 calories per day for a climatic environment with a mean annual temperature of 25° C, the planned aim is to achieve the breakthrough from the preservation-motivated society, *i.e.* to become nutritionally viable around 1973–75. The progressive raising of nutrient levels by impact of irrigation and other input factors, namely, fertilizers, mechanization, seed improvement and farm practices, is illustrated in Fig. 28.

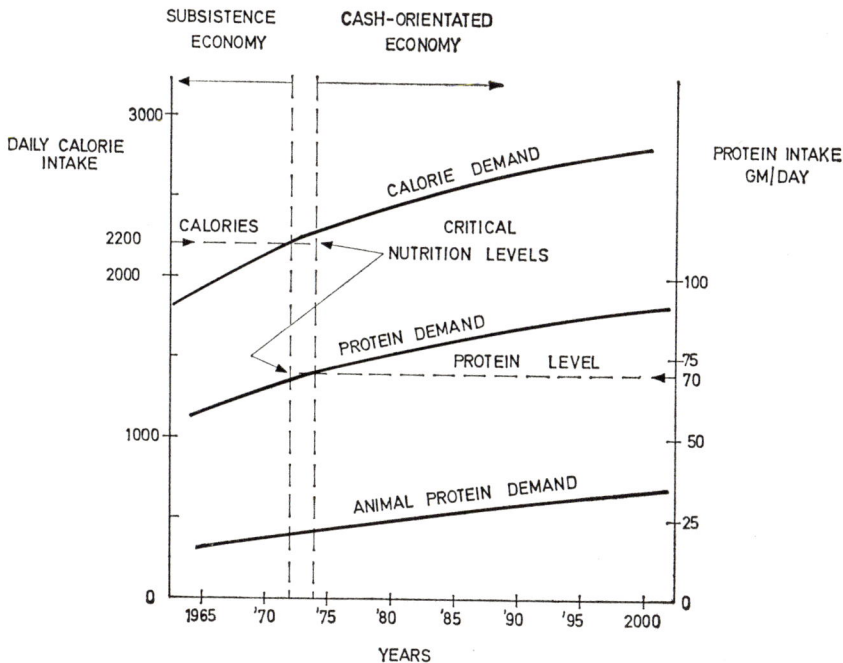

Fig. 28 West Pakistan: Projected calorie and protein demand *per capita*.

At present 75 % of the population is directly dependent on agriculture for its livelihood, providing 45 % of the gross national product in contrast to only 14 % provided by manufacturing. Agricultural products account for 78 % of the visible export earnings.

Projections of food production growth are shown in Fig. 29 which also illustrates the

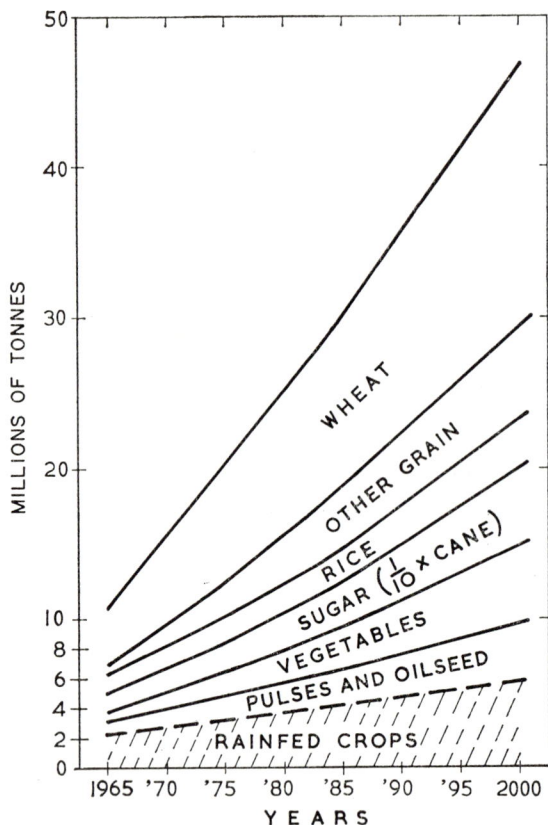

Fig. 29 West Pakistan: Projected food production.

impact of irrigation on the economy reflected in the national plan. Thus, over 70 % of the 16 million cropped hectares (40 million acres) lie within the irrigated area of the canal command and account for 80 % of West Pakistan's foodstuffs and practically all of the cash crops, whereas the extensive rain-fed agricultural areas provide only 15 % of the food production.

Projections of agricultural production growth by successive five-year plans, are shown in Fig. 30. It is seen that the planning is orientated so that the net output in relation to input costs will level out around 1980 at the relatively high ratio of 2 : 1, which compares favourably with the current South African ratio of approximately 1·4 : 1.

The response to improvements in the quantities and timing of irrigation water with

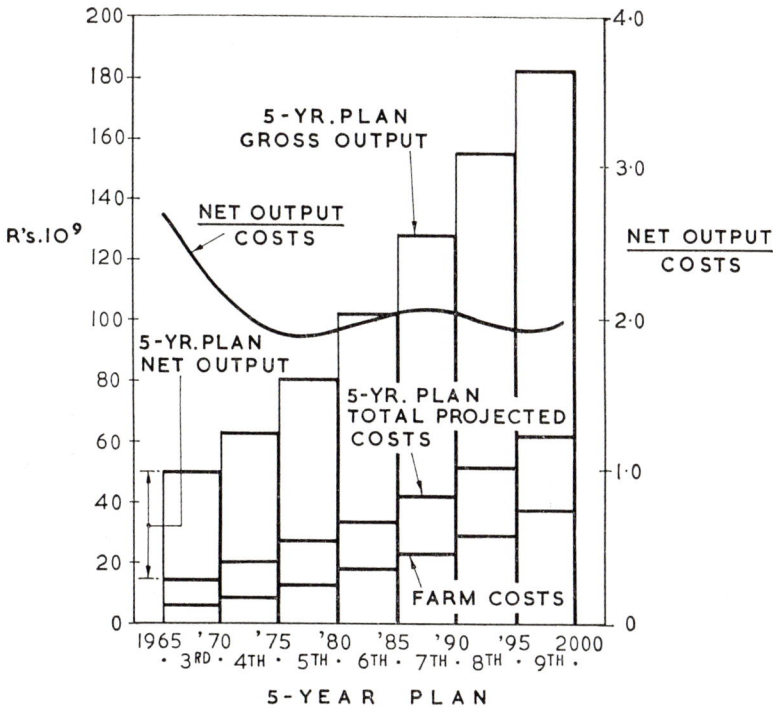

Fig. 30 West Pakistan: Projected agricultural production with farm and project costs.

respect to increased use of fertilizer and increased crop production, is shown in Fig. 31 based on records for an area where supplies from tubewells became available after 1960.

These examples of national forward planning are reinforced by two further examples of the impact of irrigation on regional economy.

Recent studies in connection with the Northern Indus plains of the Punjab have shown that, without the development programme, it will not be possible, on the basis of medium projections, to match demand with production after 1985, and that the gap will widen rapidly thereafter.

With the development programme the picture is very different. After achieving nutritional self-sufficiency about 1975, there will be a steady increase in surplus food production available for regional distribution and for export to earn foreign exchange (Fig. 32).

The other example relates to a survey carried out by agricultural economists following the commissioning in 1965 of the Trimmu–Sidhnai–Mailsi–Bahawal link system which completed the first phase of the Indus Basin project, and provided for the transfer of some 310 m³/s (11 000 cusec) of water at canal head. The impact of the link on regional economy can be seen by comparing pre- and post-link production which revealed a

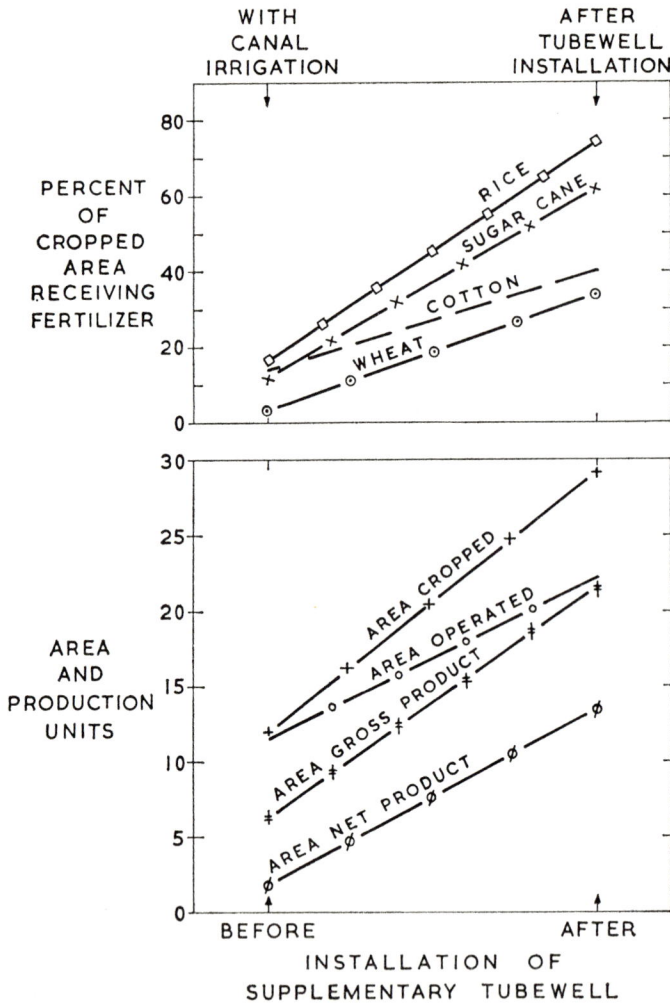

Fig. 31 West Pakistan: Impact on agricultural production following the introduction of adequate reliable irrigation water supplies after the installation of tubewells.

net product value rise in production, within one year of commissioning, amounting to £6 million.

Examples: the Columbia Basin project, Washington, U.S.A.

The second example is taken from a study carried out in a sparsely populated area in the U.S.A. where, however, consideration of nutritional viability was not a constraint.

The United States Bureau of Reclamation issued, towards the end of 1966, a report[44]

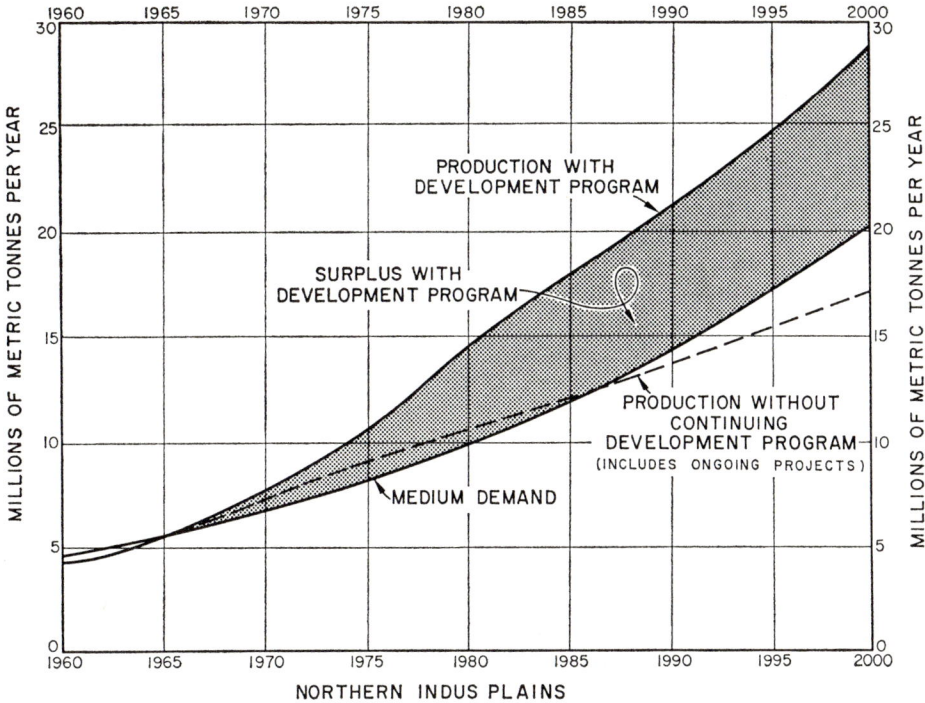

Fig. 32 Northern Indus Plains: Food production and demand 1960 to 2000.

which is unique in being the first recorded findings—based on careful studies over a period of 12 years—assessing the impact of a major irrigation enterprise, as part of a large multipurpose development, on the economic growth of the region and of the State.

Irrigation water first became available to the Columbia Basin project in 1948 and detailed studies of economic trends commenced in 1954. The study covers the 15-year period 1948–63 and is used to project conditions to 1975, when it is expected that full maturity will have been achieved with 225 000 hectares (556 000 acres) under irrigation.

The total project area comprises some 700 000 hectares, and is compared with an adjacent control (or comparison) area of 634 000 hectares of dryland farming. The relative locations of the study areas are shown in Fig. 33.

Interest in this study derives from the magnitude of this resource development exercise and the relative isolation of the project from outside economic influence. The transition from a sparsely populated, undeveloped area to an irrigated agricultural economy has created agricultural and socio-economic problems as well as providing a unique opportunity for analysing economic growth factors and trends of socio-economic development.

The areas are semi-desert, lying in the rain shadow of the Cascade mountains, with average annual rainfall varying from 202 mm in the project area to 332 mm in the comparison area.

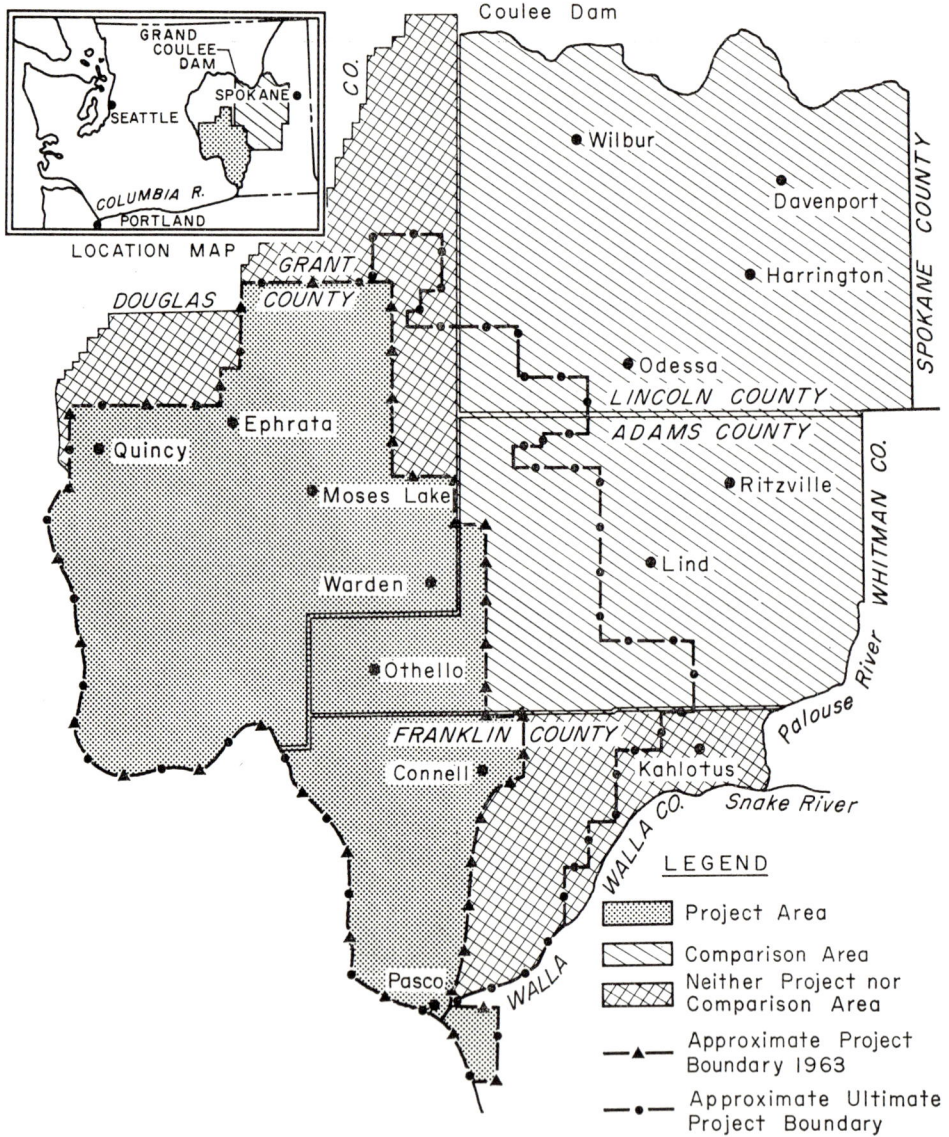

Fig. 33 Columbia basin project: Location of project and comparison area.

The key structure of the Columbia Basin project is the Grand Coulee dam with a capacity of $1·23 \times 10^9$ m³ (one million acre-feet) of which three-quarters is active storage.

The economy of the area without irrigation was extremely marginal. In order to eliminate effects arising from difference in size between the two areas, comparisons for various economic factors were made on the basis of 4000 hectare (about 10 000 acre) units.

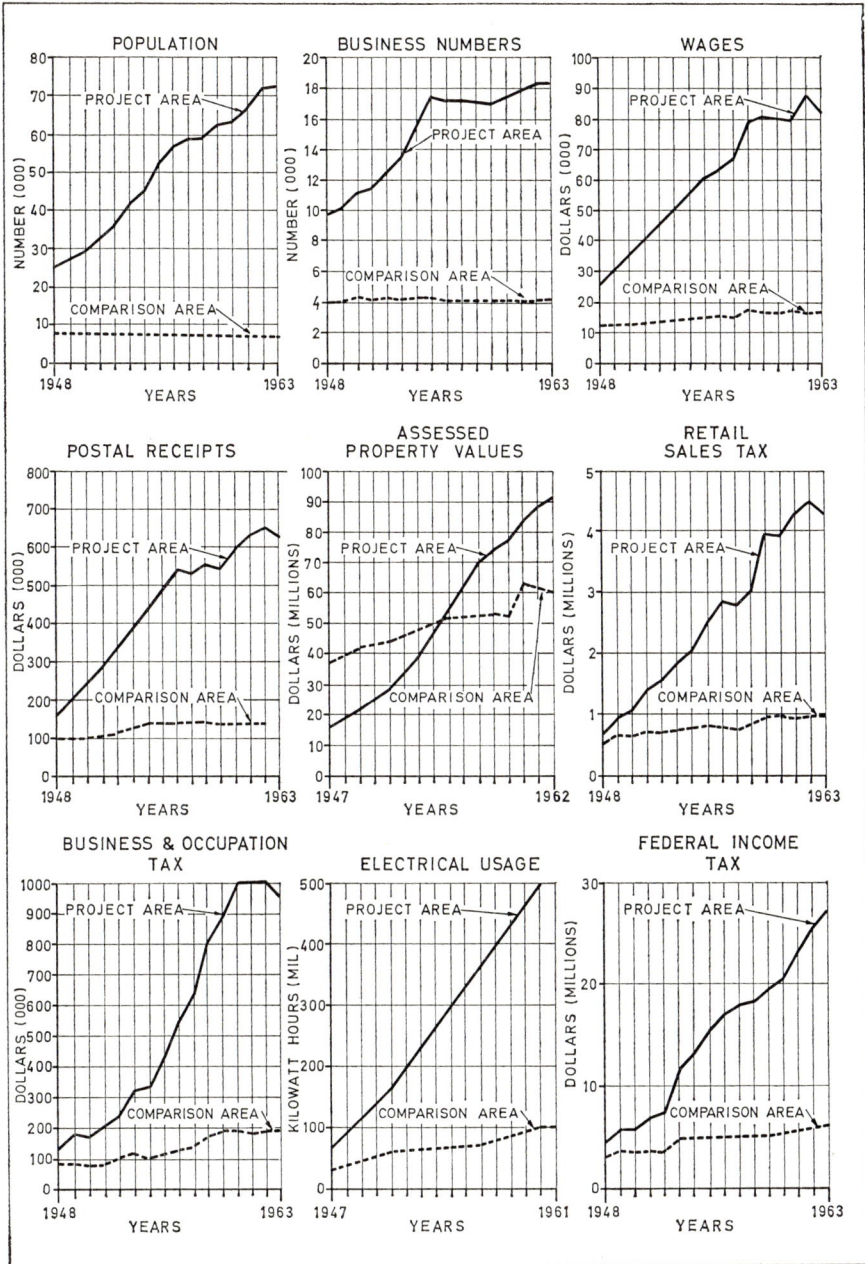

Fig. 34 Columbia basin project and comparison area: Trend of economic indicators.

Whereas it is not possible to identify and remove all the external influences on the economy, in order to obtain the pure model for the study generally, the trends of economic activity in the two areas would have been the same had it not been for the development of irrigation in the project area. Hence, it is to be expected that difference in trends since the start of the project will be due primarily to the effect of the project development.

Conclusions concerning the dynamic impact of irrigation at the end of the first 15 years of the project are shown summarized in Fig. 34 for a number of indicators. After 15 years every economic indicator for the irrigated area had grown substantially in contrast with the nominal growth of the rain-fed area, and a measure of the yeast or impact factor of irrigation is seen in the fact that the ratio of activity in the irrigated area was generally of the order of 17 times that in the rain-fed area.

(1) *Population growth (1948–1963)*
Whereas the population in the comparison area remained virtually static during the period at 16 000, the population in the project area grew as follows:

Table 26
COLUMBIA BASIN PROJECT:
POPULATION GROWTH

	1948	1963
Urban	19 160	48 240
Rural	6 000	24 000
Total:	25 160	72 240

At full maturity the project population sustained is estimated to be around 150 000 or 1·46 cultivated hectares (3·61 acres) per person.

(2) *Business numbers*
By 1963 the project area had 17 times as many business numbers per 4000 hectares as the comparison area (Table 27).

As is to be expected this ratio is comparable with that for population. Together, these economic indicators illustrate the effectiveness of an irrigation project in implementing a policy of decentralization.

(3) *Postal receipts*
Postal receipts increased from 165 000 dollars in 1948 to 635 000 dollars in 1963 compared with 104 000 and 145 000 dollars respectively for the comparison area.

(4) *Consumption of electricity*
The use of electricity is a measure of economic growth and relative standard of living.

Table 27
COLUMBIA BASIN PROJECT: BUSINESS NUMBERS

Item	Ratio project/comparison area
Contract and construction	22 : 1
Manufacturing	13 : 1
Transportation, communications and utilities	28 : 1
Wholesale and retail	14 : 1
Finance and real estate	15 : 1
Services	24 : 1
Total	17 : 1

During the period, energy consumption in the project area increased by 515% for non-farm use and by 1186% for farm users, and the comparison area increased by 228% for non-farm use and 275% for farm use.

During 1961 revenues for all purposes compared as follows:

Project area 4·16 million dollars
Comparison area 0·96 million dollars

(5) *Assessed property values*
The trends in assessed property valuation provide an indicator to relative intensity of land use as well as to economic growth.

Whereas the comparison area has shown a stable valuation base with 59% overall increase over 15 years, the project area shows 467% overall increase in the tax valuation of property.

This rate of growth, as is to be expected, is reflected in the retail sales taxes also.

(6) *Retail sales taxes*
These indicators reflect direct measurements of trends in consumer purchasing.

Thus, the average annual growth rates were: 13% for the project area; 4% for the comparison area and 7% for the State.

(7) *Business and occupation taxes*
The average annual growth rates were: project area 14%; comparison area 6%; State 9%.

The impact of irrigation development is much more apparent when comparing taxes collected on a per unit area basis. Thus, in 1963, the project area generated 105 530 dollars more retail sales taxes and 23 780 dollars more business and occupation taxes per 4000 hectares than the dryland farms. It is noticed that the ratios are again 17 : 1 and 19 : 1 in favour of the irrigation project.

(8) *Federal income tax*

Table 28
COLUMBIA BASIN PROJECT: FEDERAL
INCOME TAX

Date	Project area	Comparison area
1948	4 583 000	$3 078 000
1963	$27 169 000	$6 243 000
Total for period	$239 039 000	$75 285 000
Increase for 1948–63	493%	103%

The federal income tax payments generated per 4000 hectares of cropland in 1963 were 714 000 dollars for the project area and 41 300 dollars for the comparison area; the ratio being again approximately 17 : 1.

The higher tax payments originating from the project area clearly reflect the economic activity stimulated by irrigation development.

The impact on other areas of the State, arising from increased demand for goods manufactured, outside the region, is apparent from an analysis of rail and truck shipments. On the basis of 4000 hectares, project area inward shipments were 23 times and outward eight times greater than those of the comparison area in 1962.

By 1962, considerable differences had developed between project and comparison areas as regards growth, location and types of agricultural processing and marketing firms. Figures 35 and 36 reveal that, whereas in the comparison area little change had taken place and the activity was still mainly confined to grain silos, the project area had achieved great diversification of activity based on the 63 different types of crops grown in the area.

Other aspects of the multipurpose projects, besides irrigation, have greatly stimulated and improved the north-west economy.

Power revenues between fiscal years 1941–64 amounted to some 500 million dollars, and in different years flood control operations have averted downstream property damage ranging from 80 000 to 25 million dollars. The project area presently attracts an estimated 1·3 million days of annual recreation use. Generally, the gross revenue from power sales over 23 years represents an adequate return to pay the Treasury the total capital investment allocated to power plus 3% interest per year on the unpaid balance, together with that part of the project construction costs allocated to irrigation which is in excess of what is reasonable for project water users to pay. As regards the irrigation impact, the benefit to regional and federal economies from tax revenues alone is illustrated by the ratio of 17 : 1 for the irrigated area compared with the dryland area. Taking the excess of project over comparison areas the rate of tax revenue injection into the economy in 1964, when the irrigation scheme was only half matured, was of the order of 25 million dollars per annum. At maturity, the agricultural gross production value is expected to double

Explanation

▼ Potato packers and shippers
✕ Dry bean cleaning and storage
■ Seed pea cleaning and storage
● Onion packers and shippers
⬤ Grass and legume seed cleaning and storage
◗ Mint distillery
⬆ Pea and/or lima bean viners
▨ Feed mill

⬛ Alfalfa dehydrating and pelleting plants
■ Sugar beet refinery
⬡ Starch factory
◣ Potato and vegetable processing plants
❚ Milk distributors
⊗ Dairy processors
⊔ Meat packers
⬆ Grain elevators

▼ Number of firms handling this commodity
2 as their primary source of income

▼ Number of firms handling this commodity
(2) as a secondary or a supplemental source of income. These firms are listed at the same location according to their primary function

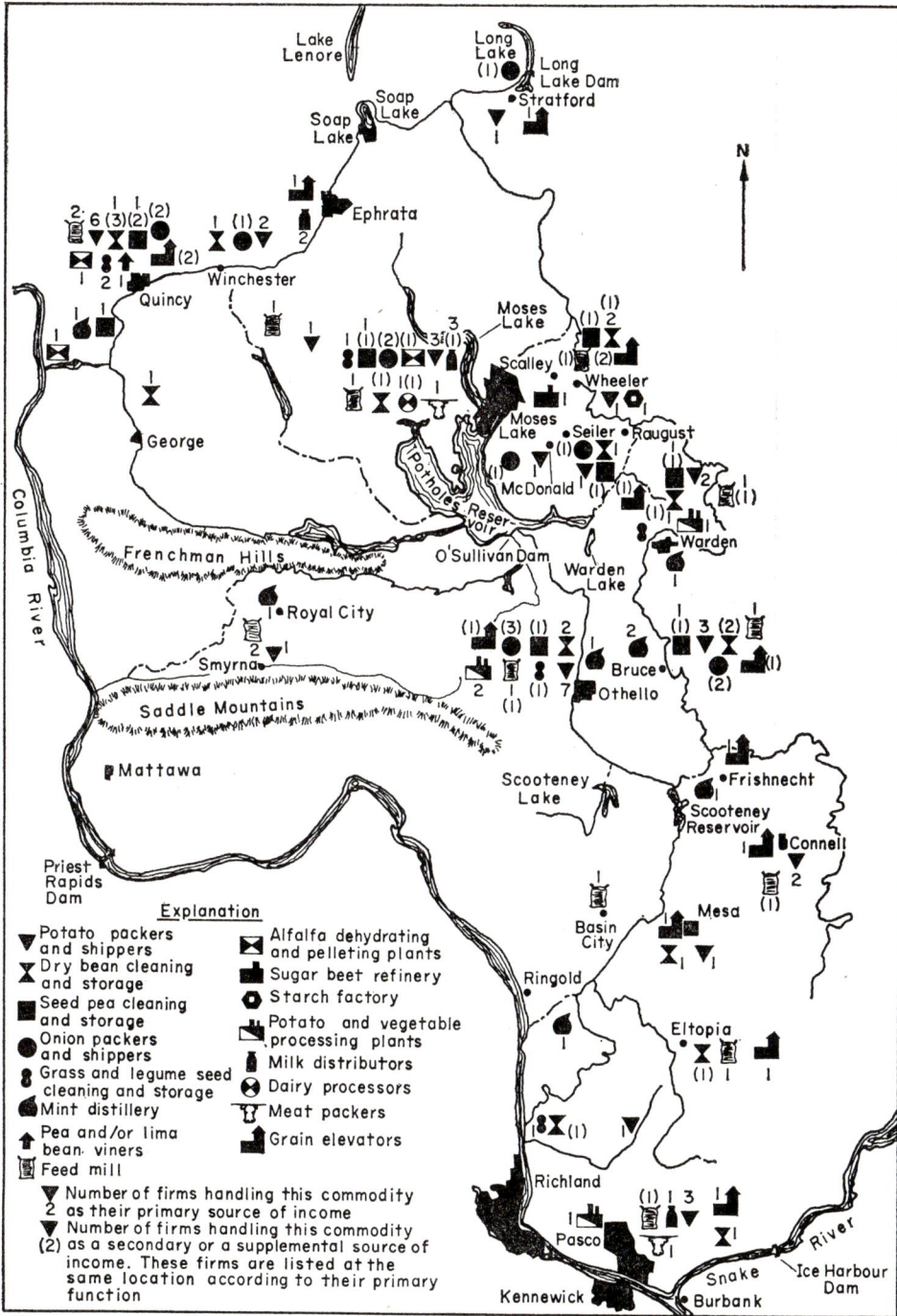

Fig. 35 Columbia basin project area: Agricultural processing and marketing firms 1962.

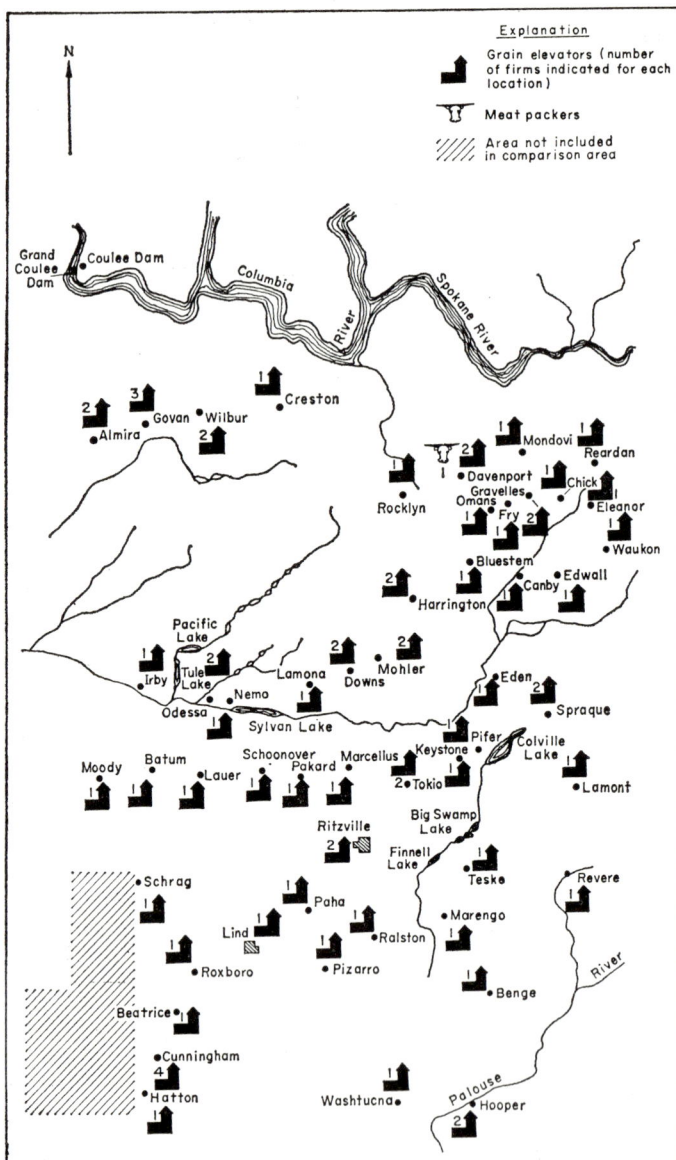

Fig. 36 Columbia basin comparison area: Agricultural processing and marketing firms 1962.

from 66 million dollars in 1964 (430 dollars per hectare or 174 dollars per acre) to 138 million dollars in 1975 (618 dollars per hectare or 250 dollars per acre) with corresponding effects on all other economic indicators. Thus, even though out of total costs the allocation to irrigation was as high as 77%, the estimates indicate that practically the whole cost of the project will, by maturity, have been returned to the Treasury by power and irrigation uses alone.

GENERAL CONCLUSIONS AND TRENDS

The following is a brief review of major conclusions and trends derived from recent studies such as the two examples quoted, which relate to widely different economic climates, and from other regions.

(a) Efficient irrigation practice generates more productive agricultural practice by stimulating the effective use of farm inputs other than water. This results in the steady generation of activity and productivity over a wide range of economic sectors. While it is not yet possible to measure the economic impact factors, the trends are becoming so clearly defined and the regional, national and international needs so pressing, that constraints to conceptual engineering arising from methodology cannot be tolerated indefinitely.

(b) Meaningful advances in engineering/economic methodology will be retarded unless attention is paid to two prime considerations:

(i) It is vital to ensure the achievement or the maintenance of nutritional viability of the communities for reasonable projections in order to be sure that economic theories based on competitive market mechanisms apply.

(ii) It is necessary to avoid confusion between terms, and in particular to distinguish between the value, the cost and the price of water. Value of water is the ideal summation of ultimate worth to the community in use and re-use of water. The cost of water is governed by decisions relating to choice of structures and location, financing and phasing of projects. The price of water is what the market will bear, economically, socially or politically. Thus, frequently, the price cannot always match the cost of the input and the relationship of both price and cost to value reflects the efficiency of utilization. This is one of the main reasons why irrigation planning and phasing cannot be dissociated from overall water resources planning and why the global trend is for central planning organizations with regional representation. Best utilization of water resources subjected to complex dynamic economic, financial, social and political pressures cannot be achieved, and its ultimate true value realized, while planning efforts persist in concentrating on random project evaluation as opposed to regional planning.

(c) As regards approach to planning, experience in the U.S.A., covering a great range of interests, has led to experiments with various integrating and co-ordinating devices. The Tennessee Valley is an example of where almost all planning and development authority was vested in a single agency leaving state and local authorities almost untouched. In the case of the Missouri and Columbia river basins, informal and special planning committees or commissions with state participation have been used. In the Arkansas–White–Red basins, New England–New York regions and the Texas and South East regions, special planning committees with federal agency and state representation have been tried. In the Delaware and Potomac basins, a single federal agency, the Corps of Engineers, has taken leadership in planning with participation of other federal and state agencies through committees. None of these arrangements have so far proved fully satisfactory or universally applicable in the U.S.A.

In Australia, with conditions more similar to those pertaining in South Africa, the trend is also unmistakably towards national planning with state participation. Having regard to population increase and the need to maintain high nutrient levels, it is estimated that shortly after 1970 Australia will be consuming most of its edible farm output unless additional farm output can be achieved. By 1990, when the population, which redoubles every 30 years, is expected to be 20 million, food production would need to be increased by 65% over 1964 levels to maintain self-sufficiency and by considerably more to benefit from developing export potential.

Variation of flow in Australian rivers is such that, on average, the volume of water required to be stored per irrigated hectare is generally twice that required in the U.S.A. Even so, in 1964 the annual gross value of production from some 1 050 000 hectares (2 600 000 acres) irrigated was the equivalent of U.S. $329 per hectare ($133 per acre) compared with U.S. $430 for the Columbia basin project, and compared with the equivalent U.S. $371 per hectare ($150 per acre) invested by the Australian Government in irrigation up to that time.

Generally, Australian experience has clearly demonstrated, first, that the ability to achieve the agricultural potential depends on many social factors, the dominant factor being the lead given by the Government and grower organizations in planning a long-term steady development, and, secondly, that the success of irrigation development depends to a large extent on the quality of research related to efficiency of water use.

(d) The formulation of long-range regional plans requires care to ensure the availability of adequate and timely water supplies. It is found that frequently more weight is given to the availability of raw materials, transport, electricity and labour than to water.

(e) In many areas, the institutional constraints, lack of education and extension services cause undue delays with respect to application of modern agricultural techniques. Because the provision of additional water supplies is not always so restricted and agricultural benefit obtained in a relatively short period, there is frequently a tendency, under pressure, to choose the water input to provide immediate impetus to agricultural yield. This is not necessarily wrong, but when applied without proper consideration of all the factors, there is scope for serious mistakes with long-term effects.

(f) Nutritional reviews, already under way in relation to water and land resources potential, may serve to indicate areas suitable for regional associations aimed at making bigger areas self-sufficient both nutritionally and economically. This is really what is behind what has been called common marketing tendencies. The concept is not confined to contiguous regions. The engineering and other disciplines involved in such planning must have regard to the rate at which transportation is being revolutionized, on land, at sea and in the air. There is little point in present planning with the tools appropriate to the ox-wagon era. As regards nutritional self-sufficiency, bearing in mind the new modes of transportation, the day may not be far off when, as in the case of monetary currencies, there are available international monetary funds, so there may become available international food banks on the basis of nutritive currencies.

6 Drainage aspect and integrated use of surface and underground water resources

A serious shortcoming frequently encountered in the planning of irrigation schemes is to limit the hydro-balance sheet to the provision of an adequate supply of water to the crops' root zone. It is evident that this first objective does not complete the full hydrological cycle. Even under efficient irrigation practice, *i.e.* allowing for proper soil leaching, there is a residual downward movement of water beyond the root zone. With inefficient irrigation—which is known to be almost universal—considerable quantities of water percolate through the soil to supplement the ground water reservoir. This movement of water involves both qualitative and quantitative considerations.

Application of irrigation water over a prolonged period of time in quantities just adequate to meet the water requirements of the crop-soil unit will result in the build up of unacceptably high concentrations of salts in the root zone of the crops. To prevent this build up it is usual to apply water at a rate in excess of net requirements so as to maintain a constant downward movement of water and salts in the soil profile. In numerous instances inadequate allowance has been made in project plans for disposal of this water, with predictable consequences.

QUANTITATIVE CONSIDERATIONS: RE-USE AND WASTE OF IRRIGATION WATER

A schematic diagram of a typical irrigation and drainage layout, based on diversion of river water, is shown in Fig. 37. This relates to a return flow study carried out for the Yakima river basin with an irrigation season extending from April to September. The fate of irrigation water diverted from the river is shown illustrated diagrammatically in Fig. 38.[16]

It is seen that of the $20\,050\,m^3$ per hectare (6·6 acre-feet per acre) diverted in the 1959–60 season:

	m³/ha	acre-ft/acre		%
Evapo-transpiration accounted for	7900	(2·60)	or	39
Loss en route to fields: evaporation, seepage and canal wastage accounted for	3950	(1·30)	or	20
Loss to sub-surface drains as return flow	4100	(1·35)	or	20·5
Loss to deep underground storage	4100	(1·35)	or	20·5

Thus available water for immediate re-use, subject to quality, amounted to 20·5%,

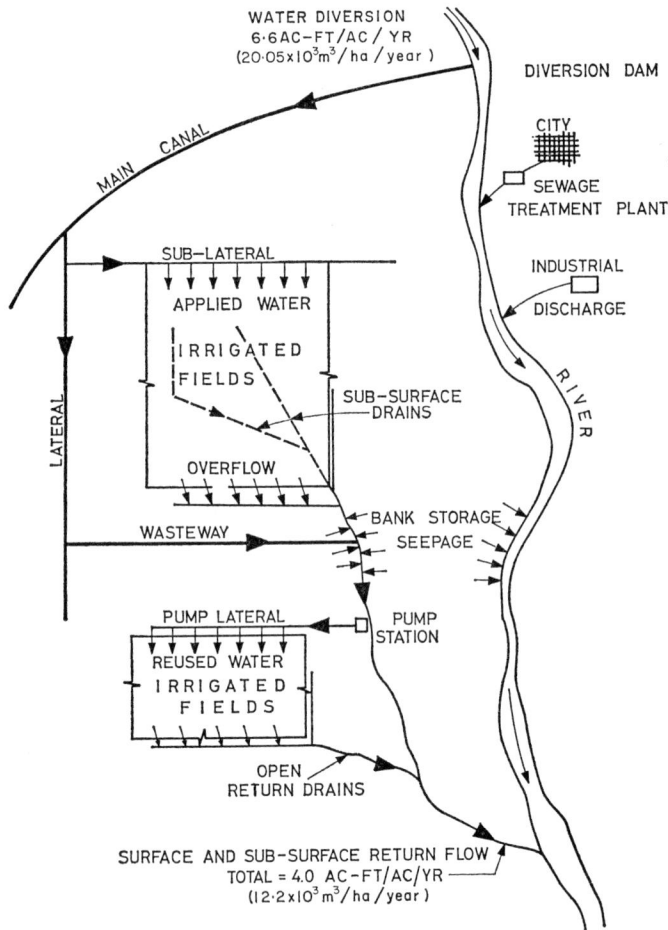

Fig. 37 Schematic irrigation supply and drainage diagram.

and the actual re-use factors from return flow during the three months July, August and September came to $3500\,m^3$, $3950\,m^3$ and $4200\,m^3$ respectively. The average quantity of return flow from sub-surface seepage during the irrigation season amounted to $685\,m^3$ per hectare (0.225 acre-feet per acre) per month.

In addition, the residual surplus water not necessarily re-used within the same year, i.e. forming an increment towards over-year storage, amounted to a further 20.5%. The effect of such deep seepage on the underground water table is indicated in Fig. 39.

In many streams in the Western United States that provide irrigation water for large areas, the summer time flow in the lower reaches of the stream may consist almost

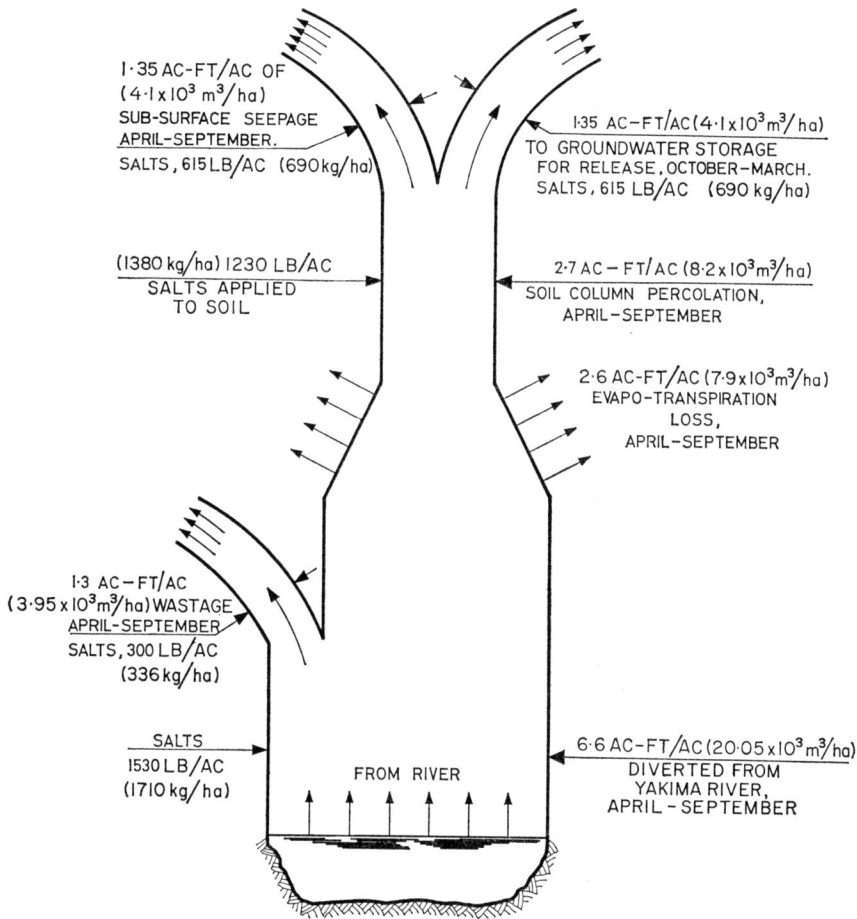

Fig. 38 Fate of diverted water and its salts: Yakima river basin 1959–60.

entirely of irrigation return flow, and in some instances water may have been used several times for irrigation.

It was concluded from this study that irrigation return flow was the major factor influencing the overall water quality of the Yakima river as compared with domestic sewage and industrial waste discharges. Leaching was the primary process responsible for the increase or change in the quantity and composition of salts in the return flow water. This is referred to further under qualitative considerations.

In territories like India and Pakistan where irrigation is practised on a vast scale and diversions are in thousands of millions of cubic metres, wastage losses of 20% together with combined deep percolation and regeneration seepage losses of 40% constitute major factors not only as regards relatively short-term economics of water/land use, but in

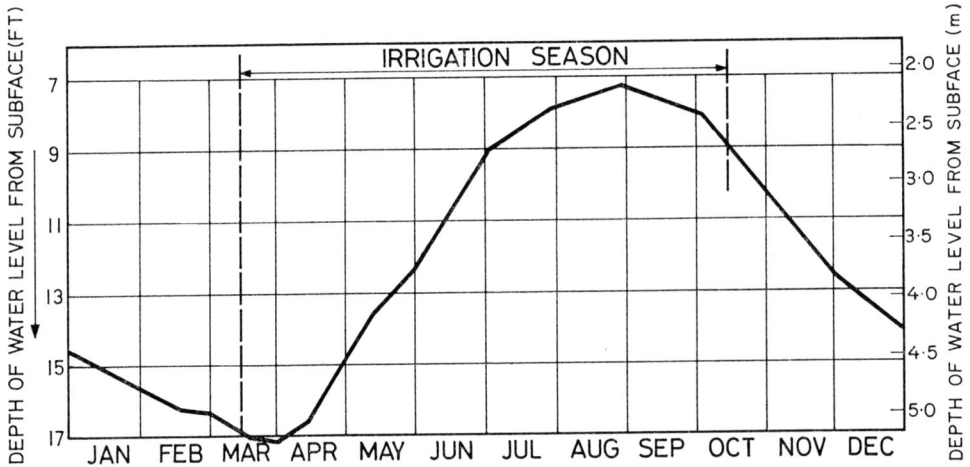

Fig. 39 Typical groundwater table fluctuation in an irrigated area of the Yakima river basin.

progressive qualitative change of waters and soils. Preliminary estimates put the annual recharge of groundwater in the Northern zone of West Pakistan at approximately $25 \times 10^9 \, \text{m}^3$ to $47 \times 10^9 \, \text{m}^3$ (20–38 million acre-feet) and in the Southern zone it is estimated to be about half this amount.

The spreading of river waters over the alluvial plains for irrigation over many years has resulted in large accumulations of underground water, with a consequent steady rise in water tables. At present over extensive areas the groundwater level has risen to, or close to, the surface, resulting in serious waterlogging and salinization.

The groundwaters of the Indus basin are highly variable in their suitability for irrigation. They are generally fresh near the rivers and canals, and tend to become more mineralized away from these sources. Where the salinity reaches moderate levels tube-well supplies can be diluted with surface supplies, but where the salinity is high, as in the central and lower part of the Indus plains, special techniques would need to be developed, such as aquifer skimming by shallow wells if this groundwater is to have commercial value for irrigation.

Such integration of surface and groundwater supplies is already being developed on a significant scale following the installation of both private and Government tubewells. In the public programme, as presently formulated, reclamation tubewells would pump an amount of water approximately equal to the long-term annual recharge. These tubewells could, however, be pumped at a greater rate during periods of shortfall in surface water supplies or in periods of extra demand. The groundwater reservoir could thus be used to balance the system supply and to some extent overcome seasonal limitations in canal capacities. (This is also referred to in Chapter 7.)

Such a concept is analogous to the procedure in electrical engineering where peak supplies, on an established curve, are provided from selected sources in accordance with efficiency rankings.

The area of the Indus plains is almost 20 million hectares (50 million acres), of which it has been estimated some 14 million hectares (35 million acres) are basically well-suited to intensive irrigated agriculture. However, because of low annual precipitation and relatively high temperatures much of the total land resources will remain unusable except for inter-mittent grazing, unless irrigation can be extended. The culturable area presently commanded by canals from the river barrages is about 13·3 million hectares. The gross sown area under canal irrigation is about 9·8 million hectares, approximately evenly divided between the kharif (April to September) and rabi (October to March) cropping seasons. This amounts to an average cropping intensity of about 70% (the sum of the area cropped in the kharif and rabi seasons expressed as a percentage of the culturable commanded area). The scope for increasing the cropped area, given an improved water supply, is thus clearly indicated.

A study of the groundwater quality zones for a particular area now in the process of being developed in the West Pakistan Punjab suggests that about 35% of the ground-water would be suitable for mixing, about 40% could be made usable subject to mixing with water of good quality and 25% would not be usable and would have to be disposed of outside the area. The quantitative significance of these figures may be assessed from the fact that in this area, covering a gross commanded area of 2·4 million hectares, of which 0·6 million hectares are under perennial and 1·2 million hectares under non-perennial irrigation, the annual groundwater recharge, including recirculation, is of the order of $10 \times 10^9 \, m^3$ (8 million acre-feet), and the accumulated reserve to be pumped out in the process of reclamation of zones of highwater table amounts to about $5 \times 10^9 \, m^3$, not all of which is suitable for irrigation.

Preliminary studies have indicated that groundwater may be classified as usable for Total Dissolved Solid (T.D.S.) Content of varying quality up to 3000 p.p.m. On this basis the following mixing ratios have been considered:

Table 29
MIXING RATIOS FOR SALINE GROUNDWATER:
WEST PAKISTAN

T.D.S. content of groundwater p.p.m.	Mixing ratios
Less than 500	No mixing necessary
500–1000	1 : 1 on an annual basis
1000–2000	1 : 1 on a continuous basis
2000–3000	1 : 2 on a continuous basis
Greater than 3000	Groundwater not usable

Additional water availability permits increased water application as well as increases in hectares cropped. In the case of the area under consideration, with balanced recharge pumping, the actual average intensity would increase from 83 % to 107 % by 1975 and the cropped hectares from 1·53 to 2·13 million (3·78 to 5·27 million acres). Concurrently with increased water supply, it is necessary to programme for increased agricultural inputs: improved supply services, with particular emphasis on fertilizer, improved seed, plant protection, and also intensive skilled extension services. Changes in cropping pattern will almost certainly result as intensity of cultivation increases.

A measure of the practical values or economic benefits foreseen from such exploitation of residual waters may be illustrated by the following estimates carried out in 1965:

On the basis of balanced recharge (*i.e.* no over-pumping) the present annual net value of crop production (at very low input values) estimated at £39 million for the particular region could be raised over the next ten years to the following levels:

Input	Net value of production £ million	Improvement per cent
Present	38·5	—
Low	58	50
Moderate	60	56
High	69	76

With overpumping the improvement could be as high as 90 %.

Moreover, it is estimated that during the subsequent 50 years the further increase in productivity would be 86 % at low inputs and about 235 % at moderate inputs.

What can be achieved for one area is clearly possible in others. The foregoing is a practical indication of the steps which may be necessary as regards skilful use of 'residual' waters in a country where, on nutritive assessment alone, it will be necessary to double the total irrigated areas by the year 2000 in order to lift the national daily diet to appropriate levels. The lessons are twofold: firstly specific regions can make a profound contribution to national averages, and secondly it is essential not to equate automatically irrigated acreages with *annual* diversions.

Another means of utilizing an underground aquifer for storage would be to increase the recharge by water spreading. In most countries this will need further research in order to demonstrate convincingly to departmental administrators the economic advantages which are not readily appreciated by politicians in a competitive atmosphere where the short-term financial advantages of certain forms of water use appear to outweigh these long-term economic advantages.

QUALITATIVE CONSIDERATIONS

It is not always appreciated that pollution by irrigation may be, and often is, progressive. Thus regular cyclic consumptive use tends to devalue the quality of remaining supplies.

Plate 5 Trimmu/Sidhnai canal RD 112: killa-bush spurs as constructed May 1st, 1965.
By courtesy of Mr. Olin Kalmbach of Tipton and Kalmbach, Denver, Colorado, U.S.A.

Plate 6 Trimmu/Sidhnai canal RD 112: killa-bush spurs berm formation by June 10th, 1965, six weeks after commissioning.
By courtesy of Mr. Olin Kalmbach of Tipton and Kalmbach, Denver, Colorado, U.S.A.

Plate 7 Backfilling with a 50 mm layer of river sand over the Polythene membrane during the construction of a reservoir in northern Nigeria.   By courtesy of British Visqueen Limited

Plate 8 One of the many large reservoirs (75 000 m³) successfully lined with Polythene sheeting in southern Spain. Constructed in 1963 to irrigate a citrus farm at Benijofar, near Alicante, it has been in continuous use ever since.   By courtesy of British Visqueen Limited

Plate 9 Canal in Tucumari, New Mexico, being lined with prefabricated butyl rubber sheets.

By courtesy of Esso Chemical Limited

Plate 10 The 'Hopalong' system being used to irrigate coffee in Kenya. Only one sprinkler line pipe has to be moved per day, 5 to 8 hours later each sprinkler is 'hopped' to the next operating position by one man, while pumping continues.

The sprinklers are mounted on risers supported by foot-operated stand-pipe couplers, equipped with automatic cut-off valves which operate when the stand pipe is withdrawn.

By courtesy of Wright Rain Limited

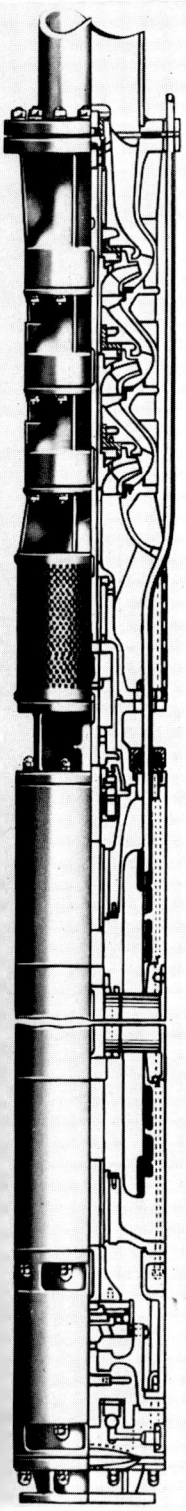

Rising main

Pump stage

Suction inlet

Motor

Motor

Plate 11   A submersible pumping installation, the 'Ulectriglide', used for harnessing underground water resources. The minimum borehole size is 250 mm and minimum capacity is 20 m³/h. Two-pole units range from 20 h.p. to 130 h.p., although up to 600 h.p. and 3·3 kV is available from four- and six-pole units.

By courtesy of the Harland Engineering Company Limited

Thrust bearing

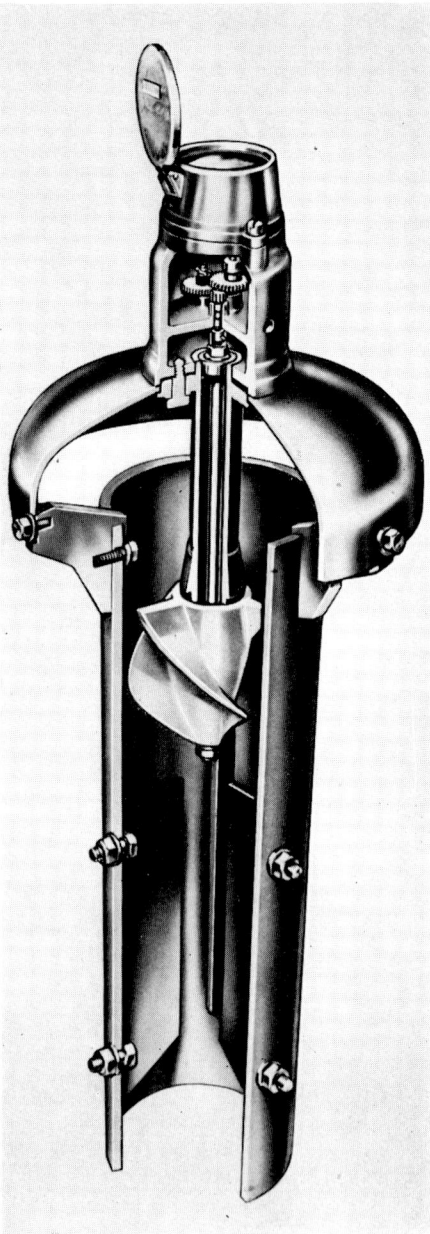

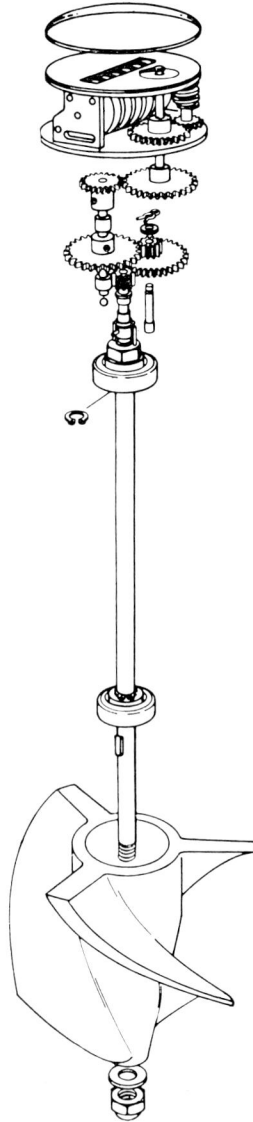

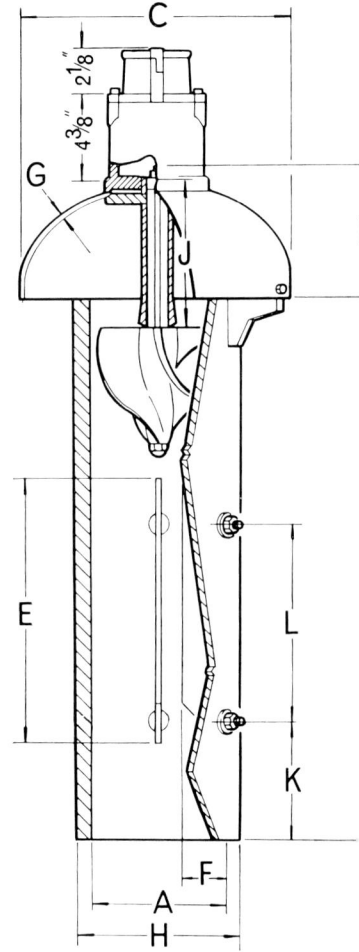

Plate 12 The 'Irrimeter', a simple free-flowing vertical flow meter. Water flows in an unrestricted, upward direction through the tube, past the propeller, and is deflected downward by the bonnet to the outside of the tube, with low head loss. The register is located on top of the waterhead assembly.

Illustrations and tabular matter supplied by courtesy of the
Hersey-Sparling Meter Company Limited

NOMINAL METER SIZE		FLOW RANGE					
		minimum		maximum			
inches	mm	gal/min	l/s	gal/min	l/s	gal/min	l/s
6	150	100	7·58	750	56·85	1100	83·38
8	200	110	8·34	1000	75·80	1500	113·70
10	250	150	11·37	1300	98·54	1900	144·02
12	300	160	12·13	2100	159·18	3100	234·98
14	350	230	17·43	2500	189·50	3700	280·46
16	400	250	18·95	2900	219·82	4300	325·94

DIMENSIONS—SPARKLING IRRIMETERS

A		B		C		D		E		F		G		H		J		K		L		Gross Weight	
in	mm	in	mm	in	mm	in	mm	in	mm	in	mm	in	mm	in	mm	in	mm	in	mm	in	mm	lb	kg
6	150	24	610	12	305	4	102	12	305	2	51	1/8	3	6 3/4	171	5 1/4	133	5	127	8 1/2	216	35	16
8	200	26	660	14	356	4 1/2	114	13	330	2 5/8	67	1/8	3	8 3/4	224	5 1/4	133	5	127	8 1/2	216	45	21
10	250	26	660	18	457	6	152	14	356	3 1/4	84	1/8	3	10 3/4	275	6 3/4	171	5	127	8 1/2	216	70	32
12	300	26	660	20	508	6 1/2	165	15	381	4	102	3/16	5	12 7/8	328	6 3/4	171	5	127	8 1/2	216	110	50
14	350	39	990	20	508	6 1/2	165	20	508	4 3/4	121	3/16	5	14 7/8	379	6 3/4	171	8	203	16	406	150	68
16	400	39	990	24	610	7 1/2	190	24	610	5 1/2	140	3/16	5	17	432	8 1/4	210	8	203	16	406	165	75

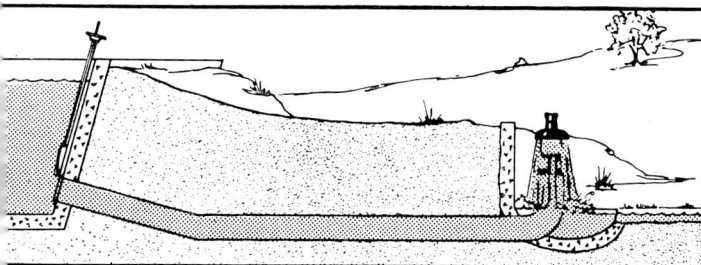

DIRECT FIELD FLOODING

An irrimeter installation requires only slight modification to an existing pipeline and the addition of an easily built structure. A rough concrete catch basin of any design, preventing soil erosion from flowing water, is adequate. Fields may be flooded or water diverted to open ditches from a typical installation.

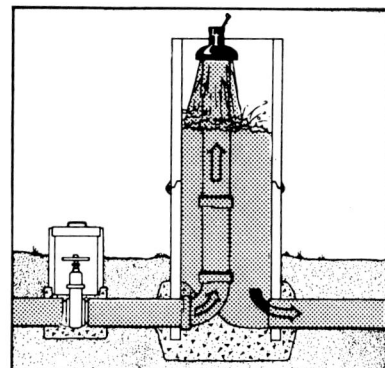

Plate 13 The effectiveness of silt trapping after two years of operation: Sipiali dam, West Pakistan.

Plate 14 Lush growth of vegetation on trapped silt after two years of operation: Sipiali dam, West Pakistan.

Growing crops use essentially pure water and leave most of the solutes in the return flows that drain from the field and eventually reappear as regenerated waters in the streams, or come to rest in the deep underground storage reservoirs, In addition, the water applied to irrigated fields increases the weathering rate and leaches salts from the soils. This phenomenon is illustrated in Fig. 7, Chapter 1, which records a sevenfold increase in salt concentration of the stream as a result of irrigation on one project in the U.S.A. Half of the increase in concentration resulted from evapo-transpiration (crop-use) and half from leaching.

On the Yakima experiment relating to 60 km (37 mile) long canals, partly lined and partly unlined, with an average flow of 27 m^3/s (980 cusec), it was found from 47 sampling stations that in general water quality changes in the main irrigation canals were of minor importance compared with quality changes which occur as a result of waters applied to the soil being irrigated. These consisted of fine sandy and silt loams of variable depths underlaid with gravel or basalt. By calculation, evapo-transpiration loss alone would account for a multiple of 1·7 of the original salt concentration.

The actual experience was that application of a low salinity water resulted in marked chemical quality changes to the applied water. The change in anion and cation concentration between the point of application and sub-surface return flow was found to be between 5 and 6 fold. A tenfold increase in nitrates and a threefold increase in phosphates indicated a possible loss of applied fertilizer.

Reported changes of particular interest were a marked decrease in

Temperature	dissolved oxygen
Turbidity	colour
Coliform bacteria	

and a large increase in

Bicarbonate alkalinity; hardness
conductivity; chlorides
nitrates; phosphates
sulphates and principal cations

It was apparent that leaching, and to a lesser extent, ion exchange, were responsible for the changes in quality of the return flow in this area. For every 400 hectares of land under irrigation an average of 83·5 tonnes of salts were removed from the soil per month primarily by leaching. The majority of the leached ions consisted of bicarbonates, calcium sulphate and sodium.

It was also found that, as compared with domestic sewage and industrial waste discharges, irrigation return flow was the major factor influencing the overall water quality of the Yakima river, and that leaching was the primary process responsible for the change in salt content of return flow.

Sub-surface drainage on entering an open drain during the irrigation season was

E

modified in quality by the influx of surface run-off from over-irrigation and canal wastage. Sub-surface drains continued to drain throughout the non-irrigation season as the water table was drawn down.

It was concluded that the quality of return flow would be significantly improved if more care were exercised in handling of canal wastage and if over-irrigation were reduced, and if a greater proportion of closed drains were used.

As expected, it was found that the method of water application and the type of crops had no marked effect on the return flow.

High concentrations of salts in return flow are not necessarily deleterious to downstream users unless contained in large quantities of water. The Yakima data shows that whereas tonnage of salt per month per m^3/s was greatest during the irrigation season, the total salts in relation to total flow showed just the reverse because of the reduced flow of the river during the irrigation season.

The down project river water with its return flow would be classed as a low salinity water suitable for irrigation re-use.

A further example of the change in quality of irrigation water is given in Table 30, which compares the composition of irrigation water and the drainage water on the Vaalhartz irrigation scheme in South Africa.[45]

Table 30

COMPARISON BETWEEN THE COMPOSITION OF
INTAKE WATER AND RETURN DRAINAGE FROM THE
VAALHARTZ IRRIGATION SCHEME
(figures in parts per million)

Constituent	Intake water	Return drainage
Total dissolved solids	95	330
Total alkalinity (as CaCO₃)	45	125
Total hardness (as CaCO₃)	55	169
Magnesium hardness (as CaCO₃)	31	83
Calcium hardness (as CaCO₃)	24	86
Sodium (as Na)	10·3	48
Potassium (as K)	6·3	3·8
Chloride (as Cl)	12	66
Sulphate (as SO₄)	3·0	50

One factor of particular engineering significance is that large reservoirs, constructed between or above major irrigation projects, can have a marked influence on annual salt balance by retention of, and/or dilution of, salts—particularly if the geology of the reservoir is such that bank storage is an appreciable factor in its operation. In a study based on data collected over the period 1931–46 Bliss[46] found that some 12·8% of total dissolved solids in the Rio Grande were lost in storage or transit through the Elephant Butte reservoir in south central New Mexico.

Salt concentration rates would seem to vary with rising and falling stages of the reservoir. Bliss found that in the case of Elephant Butte, where approximately 25 % of water stored during the peak years 1941–43 was apparently held in bank storage, a direct relationship existed between bank storage of water and storage of the more soluble ions such as sodium plus chloride. This suggests that the apparent losses and gains of these two ions may serve as indicators to check the measurement of in and outflow of water from bank storage.

Having regard to the importance of time disciplines in irrigation and factors controlling the effectiveness of rainfall and hence of irrigation, it is important to bear in mind that chemical properties of the soils also influence the rate of infiltration. The swelling of the soil colloids is a well-known phenomenon in changing pore-size distribution, which in turn affects rate of infiltration. This swelling is particularly noticeable in the presence of exchangeable sodium.

PHREATOPHYTES

It is becoming increasingly difficult to ignore the effects of phreatophytes, marsh plants and various 'wet-foot' crops on the hydro-balance sheets of countries like the U.S.A., Africa and the Middle East. The economic impact arising from the immense water losses in the form of unproductive evapo-transpiration from these 'pests' are tremendous.

The paddy rice fields of Ceylon are practically choked with the weed 'Salvinia Auticulata' the spread of which, on Lake Kariba, was halted by natural phenomena (probably wave action) when a lake area of 388 km² (150 square miles) had become infested.

In the U.S.A. the spread of phreatophytes like saltcedar (tamarisk) and cottonwoods is estimated to cover 64 800 km² (16 million acres) of the western half of the U.S.A.—where water is already a limiting factor. Consumptive use of such water-loving plants is generally about double that of dry-foot crops. The annual non-productive loss of fresh water by this route in the Western United States is estimated at 30×10^9 m³ (25 million acre-feet). In order to maintain perspective one must evaluate the cost of producing this amount of freshwater annually by, say, desalinization processes. It has been claimed that this process is now economic, *i.e.* competitive with other sources of supply. Even at £0·10 per 4500 litres (1000 gals.) this would represent an annual outlay of nearly £25 million.

If only 10 % of the 324 000 km² of perennial swamps of Africa were to be drained and suitably irrigated, about 32 400 km² would be added to the 52 700 km² potential irrigable area in the region south of the Sahara.

Phreatophytes need not always, however, be regarded as wasting monsters! In some areas of the Middle East the use of tamarisk and cottonwoods has been considered with respect to windbreaks—sometimes for crop protection and sometimes to halt the march of moving sand dunes—especially on the edges of canals.

Researches by Basov on the forest plantations in the Kamennaya steppes of Kazakhstan in Russia[47] have demonstrated the existence of 'depression cones' in the water table underneath these plantations during the vegetative period. Kroylov has demonstrated in Uzbekistan that when the water table is near the surface phreatophytes consume considerable quantities of underground water.

This leads to consideration of the concept of using phreatophyte shelterbelts in regular patterns for the reclamation of saline soils in heavily waterlogged regions such as occur in West Pakistan.

Reclamation of such soils in flat terrain with a slope of 0·2 m per km (1 foot per mile) by means of tubewells and open drains represents a considerable financial layout. Yet the problem cannot be ignored as waterlogging is increasing at a rate of some 40 000 hectares (100 000 acres) per year. In specific instances phreatophytes may yield good results. The method would be to commence reclamation treatment on the fringes of the infection—where deterioration has not proceeded so far as to inhibit crop growth and to work towards the eye of the problem area.

In circumstances such as these it may be worth developing pilot projects consisting of square plots bordered by salt tolerant phreatophytes so spaced as to draw down the water table within each square to an extent which will permit ordinary crops to grow after or with leaching. The possible effects are illustrated diagrammatically in Fig. 40.

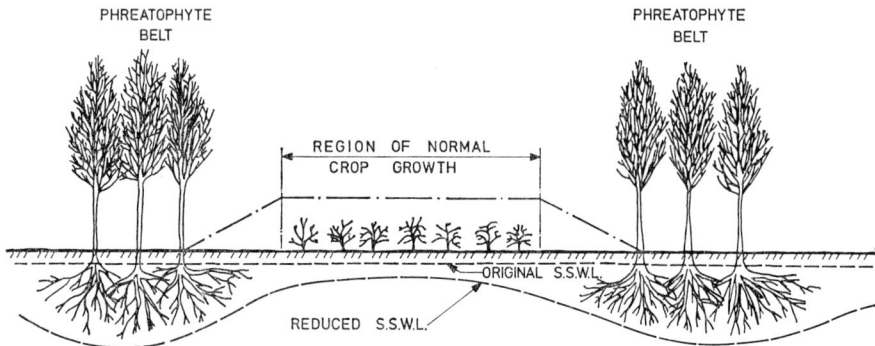

Fig. 40 Diagrammatic representation of the use of phreatophytes for soil reclamation by lowering of subsoil water level.

In addition to the direct advantages of soil reclamation and flexibility of development there would accrue as by-products the considerable advantages of shelterbelt cultivation.

Generally the advantages of windbreaks arise from the ability to reduce wind velocities to less than 19 km/h. (12 mph) which is the threshold velocity at which soil particles begin to move. Protection afforded to leeward is proportional to height.

Within a shelterbelt area the humidity is generally higher than in the open fields with corresponding effects on evapo-transpiration and evaporation. The general effect of cottonwood shelterbelts on evaporation is shown in Fig. 41.

The following typical research summaries indicate general trends:
(H = height of windbreak)

(a) *Type:* Series of parallel field windbreaks.
   Season and year: 4-year study in summer, 1950s.

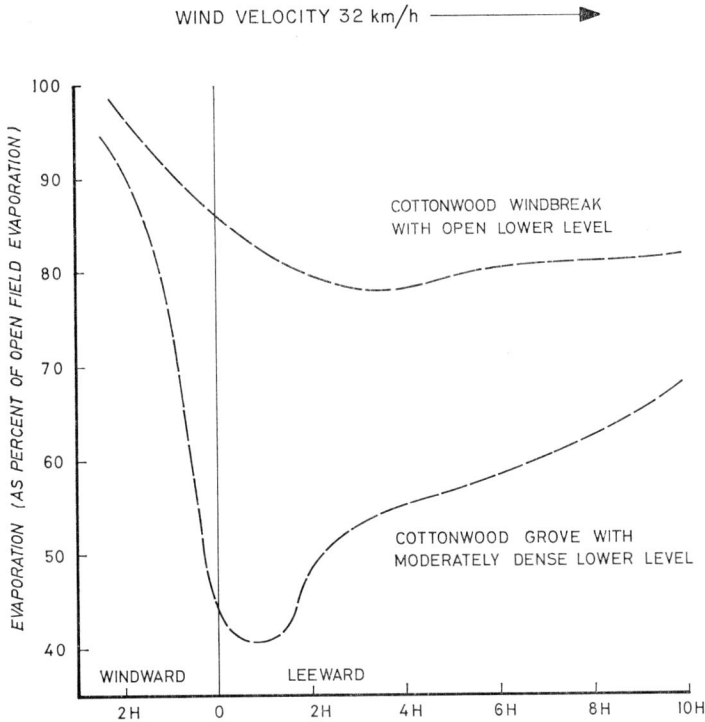

Fig. 41 Percent of open field evaporation windward and leeward of cottonwood windbreaks of different
    structures.

Place: Kamennaya Steppe, Russia.
Reference: Molchanov, 1956.[48]

Average monthly relative humidity was 2 to 3% higher during the day on oat fields
between windbreaks than on oat fields of the open steppe.

Relative humidity was 4 to 5% higher on protected fields of oats than on open steppe
oats on hot summer days.

Relative humidity was 2 to 3% higher on protected fallow fields than on open steppe
fallow fields.

(b) *Type:* Series of field windbreaks.
    Season and year: Summer, 1933.
    Place: Russia.
    Reference: Sokolova, 1937.[49]

Average relative humidity was 8% higher in fields between windbreaks than on open prairie.

Relative humidities average 11% higher between windbreaks than on open prairie in the *morning hours.*

(c) *Type:* (Not stated)
   Season and year: Summer, 1940s.
   Place: Germany.
   Reference: Steubing, 1952.[50]

Dewfall was 200% greater on fields protected by windbreaks than on open fields. Heaviest dew was in the 2 to 3 H leeward zone.

(d) *Type:* Windbreak of Japanese Black Pine, 1·5 m tall and 6 m wide.
   Year: 1940s.
   Place: Japan.
   Reference: Iizuka and others, 1950.[51]

Evaporation rates leeward of windbreak compared to open field were:
40% at 5 H
60% at 10 H
80% at 20 H
Same as open at 25 H.

(e) *Type:* Series of parallel 3-row, 6 m-tall windbreaks at 20-m intervals.
   Season and year: May–September 1951–54.
   Place: Saskatchewan, Canada.
   Reference: Staple and Lehane, 1955.[52]

Evaporation from water surface was 15% less at 4 H leeward and 9% less at 16H leeward than in open.

(f) *Type:* Field windbreaks at right angle to slope.
   Year: 1930s.
   Place: Kamennaya Steppe, Russia.
   Reference: Basov, 1941.[47]

Runoff on an area with 6% of land in field windbreaks was reduced to 43 from 63% compared to runoff on open steppe. With 18% of land in field windbreaks runoff was reduced to 25%.

The effects on crops within such shelterbelt areas are indicated by the following:

(i) *Type:* 40-year-old cottonwood and boxelder windbreaks of medium density, and averaging 12 m tall.
    Year: 1935–41.
    Place: North Dakota, South Dakota and Nebraska, U.S.A.
    Reference: Stoeckeler, 1962.[17]

Wheat, rye, barley, and oat yields showed a 36-bushel total increase per 0·8 km (half a mile) of windbreak in the 0 to 14 H leeward zone of high-yielding fields over the average of unprotected areas, in North Dakota and South Dakota.

The same crops showed a 74-bushel total increase per 0·8 km (half a mile) of windbreaks in the 0 to 14 H leeward zone of low-yielding fields over the average of unprotected areas.

Corn yields averaged 19 % greater in the 2 to 10 H leeward zone east of windbreaks than on unprotected fields in Nebraska.

(ii) *Type:* System of field windbreaks of various types.
    Year: 1930s.
    Place: Juibyshev, Russia.
    Reference: Karuzin, 1936 and 1947.[53]

Barley, oats, and winter wheat in shelter of windbreaks compared to open unsheltered fields gave yield increases of:
    (a) 100 to 400 % in severe drought years.
    (b) 50 to 60 % in moderate drought years.
    (c) 10 to 15 % in years of no drought.

(iii) *Type:* Tree windbreaks and artificial barriers
    Year: 1935.
    Place: North Dakota, U.S.A.
    Reference: Bates, 1944.[54]

Alfalfa yielded 60 to 70 % more in best part of protected field as compared to overall field average.

(iv) *Type:* Willow–ash windbreak 4 m tall.
    Year: 1950s.
    Place: Japan.
    Reference: Matsui and Yokoyama, 1955.[55]

Rice yields decreased 51 % at $\frac{1}{2}$ H leeward, but were increased 3 % at 1 H; 33 % at 3 H; 49 % at 6 H; 33 % at 9 H; 28 % at 12 H; and 8 % at 15 H, compared to open, unprotected fields.

(v) *Type:* 4-row deciduous tree windbreak, 6 m tall.
   Year: 1919–22.
   Place: Jutland, Denmark.
   Reference: Soegaard, 1954.[56]

Apple yields in bushels per hectare in sheltered area of windbreak were:
(a) 432 at 1·5 H
(b) 222 at 3 H
(c) 97 at 5 H
(d) 74 at 6 H
(e) 64 at 8 H

A general diagrammatic representation of cropyield in relation to leeward distance from a windbreak (after Stoeckeler) is shown in Fig. 8 in Chapter 1.

7 Efficiency of water distribution and use on the land

It must be emphasized at the outset that responsibility for the solution of the complex dynamic problems associated with irrigated agriculture cannot be the prerogative of one or two disciplines and that the closest co-operation between a large number of interested and qualified authorities is essential.

In the purely engineering field, there will, of course, remain the task of creating new storage facilities, despite such problems as scarcity of suitable dam and reservoir sites, irregular geographical distribution of precipitation and runoff, and mounting costs of labour and materials. In the long term, the aim must be to shorten construction periods and reduce capital costs by the introduction of new materials and/or new methods of construction. This is bound to be a slow process, having regard to the numerous complex problems involved.

Regions with poor hydro-meteorological characteristics carry an initial handicap as regards costs of water supply. In this respect South Africa is particularly penalized owing to the relatively higher storage capacity required to yield firm flows comparable with those of other countries. This is illustrated in Figs. 42 and 43 based on computer studies for a recurrence interval of 50 years.

The comparison of storage capacity required (as a percentage of the mean annual run-off) to ensure a safe yield of 80% of mean annual run-off may be summarized as follows:

Vaal river (South Africa)	200%
Three Indus rivers (West Pakistan)	40%
The Blue and White Niles (Egypt/Sudan)	10–40%
The Tigris and Euphrates (Iraq)	80%
The Australian rivers	100%

Thus, compared with the Australian river selected, the Vaal needs *twice* the storage capacity for a given assumed yield, and *five* times that required for the Indus rivers.

This question is further highlighted by a comparison of these characteristics for different rivers in the Republic, as shown in Fig. 44, which illustrates how valuable the basin characteristics of the Upper Orange and Tugela rivers are, in terms of storage efficiency, compared with the Vaal and Riet rivers.

However, as a result of easier communications and a more generous international dissemination of pertinent statistics, it has become possible in recent times to make re-appraisals of existing practices. There have emerged, largely as by-products from studies concerned with the effects of malpractices in irrigation, some new concepts and lines of approach which are of interest to all concerned with the beneficial development of water and land resources including the financial promoters, but particularly, in the first place,

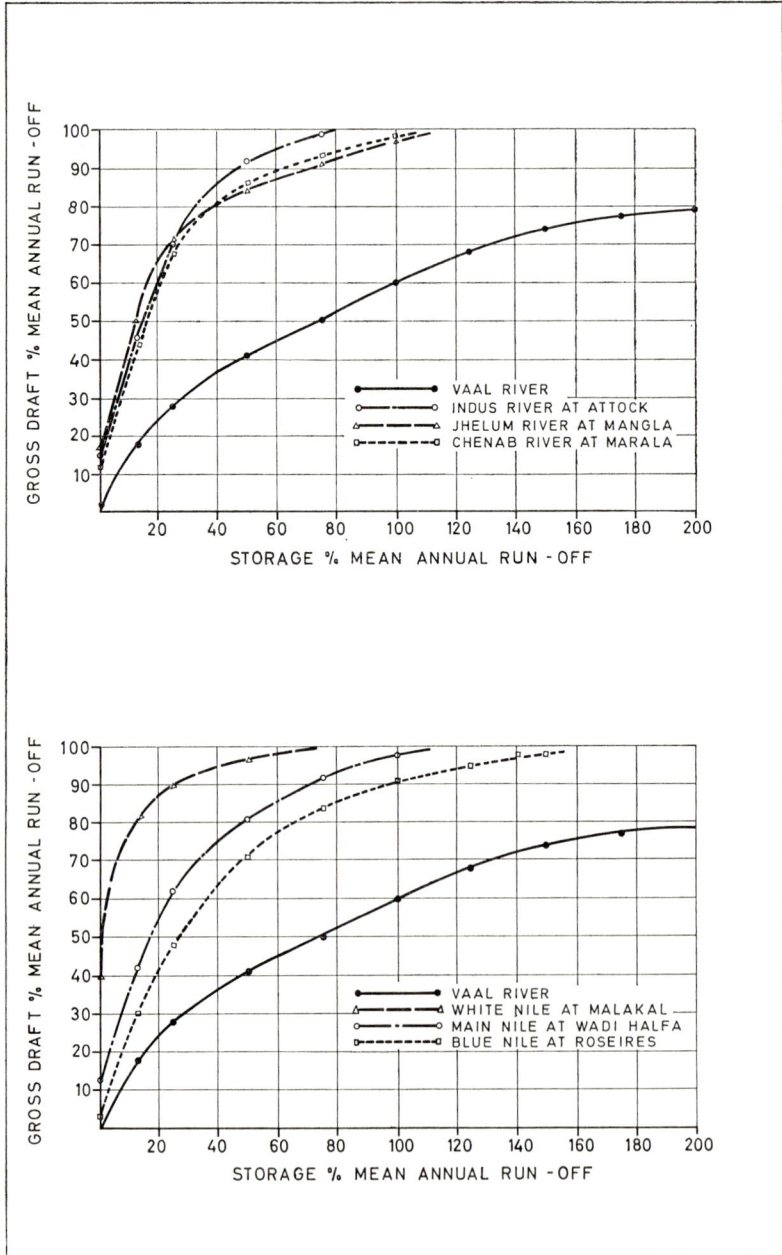

Fig. 42 Comparison of storage-gross draft curves: 50-year recurrence interval.

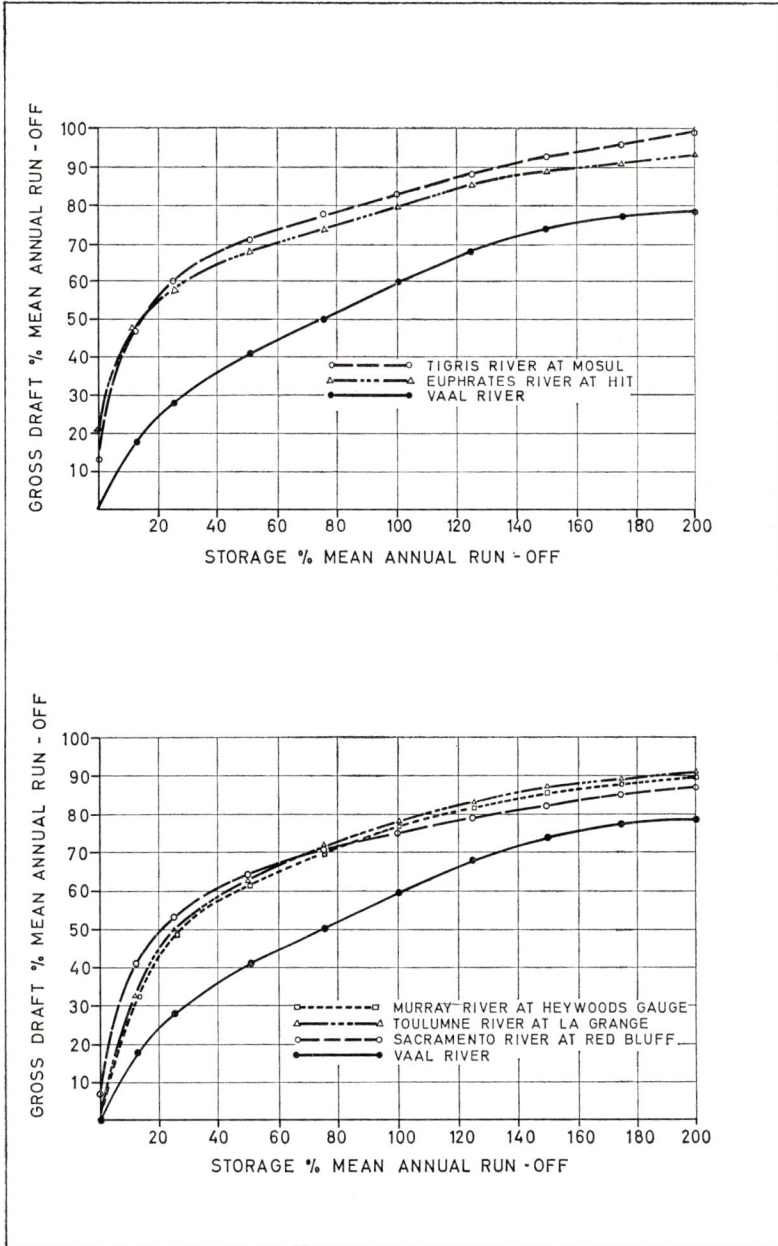

Fig. 43 Comparison of storage-gross draft curves: 50-year recurrence interval.

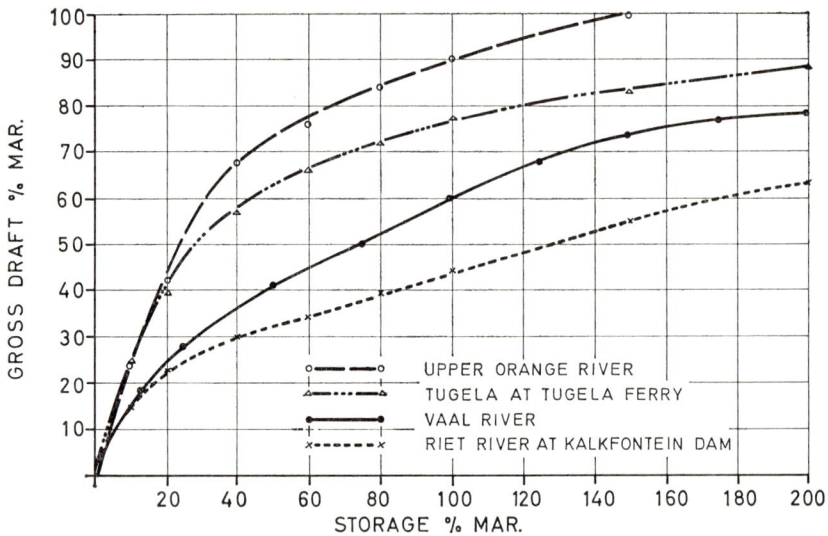

Fig. 44 Comparison of storage-gross draft curves; 50-year recurrence interval.

to the agriculturist and the engineer. It is becoming apparent that, as a result of more effective conveyance, distribution and use of existing water resources, agricultural output may in some cases be more than doubled.

BASIC OBJECTIVES OF IRRIGATION

The basic purpose of irrigation is to increase crop production per unit of land area, and the objective of efficient irrigation must be to obtain the highest possible sustained agricultural return per unit of water supplied from source. Such returns may be in the form of market values of cash crops, food values of subsistence crops, or feed value of forage crops for livestock.

The demands for water from competitive interests—domestic, industrial and agri-cultural—are inevitably expanding to the extent that available supplies are becoming restricted in some densely populated regions. The unit value placed on water, whether as a utility or a necessity for generating wealth, will tend to feature increasingly in the allocation of water resources. The average efficiency of the use of water resources in irrigated crop production varies over wide ranges, and is generally low at present. In consequence, the unit value of irrigation water at source is also low.

This shortcoming will increasingly become a problem for agricultural interests in both arid and humid regions, and the only foreseeable means of increasing the value of irriga-tion water is by improving its effective use in the production of crops.

In humid regions, the careful timing and proportioning of supplemental irrigation water to make up short-term deficiencies of precipitation may revolutionize farming, not merely as regards gross increases in output, but particularly with respect to stabilization of yearly returns.

In all these circumstances, the trend is for new projects to come under increasingly close scrutiny for economic and financial feasibility. There would thus seem to be a case for consultation with agricultural economists and financial experts during the early stages of project appraisal for purposes of co-ordinating with agricultural and engineering aspects.

IRRIGATION EFFICIENCY

Project Irrigation Efficiency (E_i) is usually defined in terms of the following expression:

$$E_i = \frac{W_c}{W_r} \times 100 \text{ (per cent)}$$

where

W_c = the quantity of irrigation water actually consumed by crops in the project area over a given period,

and

W_r = the quantity of irrigation releases from source over the corresponding period, whether from surface or groundwater storage or run-of-river diversions.

This definition may be developed further as follows:

$$E_i = E_a \times E_c$$

where

E_a = Application Efficiency

$$= \frac{W_c}{W_d} \times 100 \text{ (per cent)}$$

in which W_c is as defined above,

and

W_d = the quantity of irrigation deliveries to farms or fields over the corresponding period,

and where

E_c = Conveyance Efficiency

$$= \frac{W_d}{W_r} \times 100 \text{ (per cent)}$$

in which W_d and W_r are as defined above.

These definitions serve to classify responsibilities for effecting improvements as between the individual cultivator and the administrative authority for the project. The application efficiency (E_a) is largely the concern of the cultivators, being influenced by the manner and timing of water applications, field configuration and slopes, soil characteristics, types of crop under cultivation and the cultural practices employed. The conveyance efficiency (E_c) is mainly the concern of the authority, relating as it does to losses and defects in the conveyance system, and the timing of releases.

Special considerations

It is self-evident that the quantity and timing of releases into conveyance systems should be dependent on total actual application requirements. It is essential, therefore, that prior considerations in all studies should be given to actual crop requirements.

As a result of recent researches, it is now possible to prepare average or extreme (dry year) characteristic curves of basic irrigation requirements for a particular locality, calculated from simple climatological data for the area. (See Chapter 2.) Figure 45 illustrates a typical set of climatological data and the derived basic average irrigation characteristic curve for Hyderabad in West Pakistan. After suitable adjustments, based on experienced judgement, to allow for application and conveyance efficiencies, the resultant set of curves may be used to assess maximum and minimum quantities and intervals for irrigation application for purposes of designing storage, conveyance and distribution systems. The method is illustrated in Fig. 46.

Subject to regular accountancy as described later and periodic adjustment for deviations from these primary (design) characteristic curves due to climatological deviations from those adopted, a secondary set of curves may be deduced for operating purposes for an individual growing period or year. These may be used as controls for actual applications, and hence to determine withdrawals from source.

Factors for consideration in the selection of application depth are soil texture and the rooting depths of crops. It is, therefore, useful to have an idea of the depth of applications required for various soil textures and root depths to fill the corresponding effective storage volume from wilting point to field capacity. Such a set of curves for well drained soils is shown in Fig. 47. It is, of course, necessary to make allowance for the actual state of moisture of the soil in question.

The indications are that one of the major contributory factors to waterlogging, particularly in regions where large irrigation canals have been constructed, may be the regular application of excessive water out of phase with ideal requirements dictated by climatological, soil and root zone depth factors. Some typical examples of incorrect phasing are illustrated diagrammatically in Fig. 48.

Application efficiency

Attempts to come to grips with the problem of increasing application efficiencies at the fields reveal that the expression:

Hyderabad, West Pakistan, 25° 23'N 68° 25'E

Temperature and relative humidity

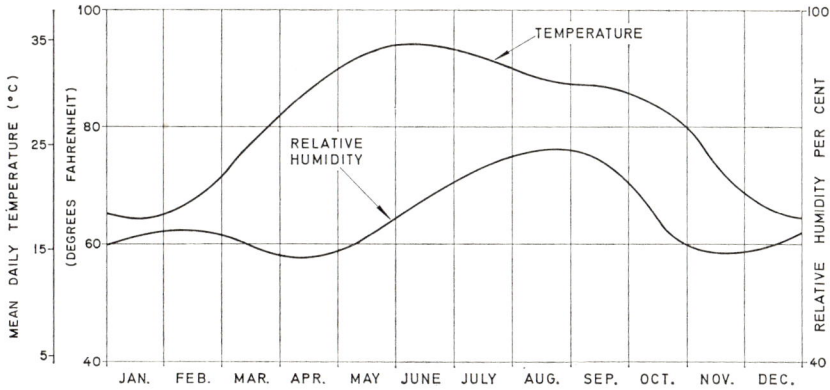

Crop consumptive use and effective rainfall

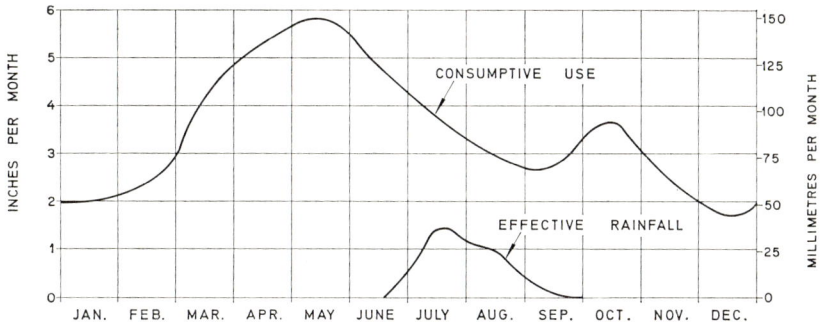

Cumulative basic irrigation requirement (average year)

Fig. 45 Climate and irrigation characteristic curves for Hyderabad (West Pakistan).

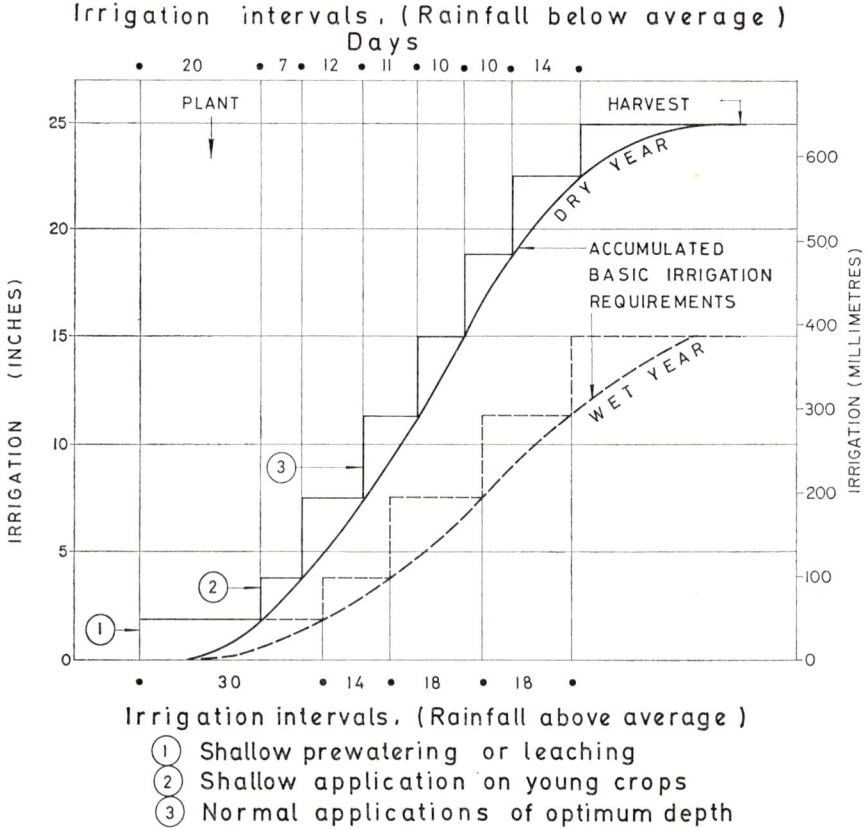

Fig. 46 Schematic irrigation requirements and field application.

$$E_a = \frac{W_c}{W_d} \times 100 \text{ (per cent)}$$

is an oversimplification in the sense that it omits reference to vital factors affecting efficiencies, and thereby tends to disguise inefficient operations, and to hamper constructive lines of approach towards achievement of greater economy in the use of water, land and money.

As a first approach, it is reasonable to subdivide the efficiency ratio into three constituent factors as follows:

(a) *A supply/demand or diversity factor*
For any given locality, there exists a characteristic climatic stimulus controlling evapotranspiration. This stimulus exists whether or not there is available water to be transpired and evaporated and/or crops. To the extent that for any particular time period, actual supplies available to the crop significantly exceed or fall below the corresponding

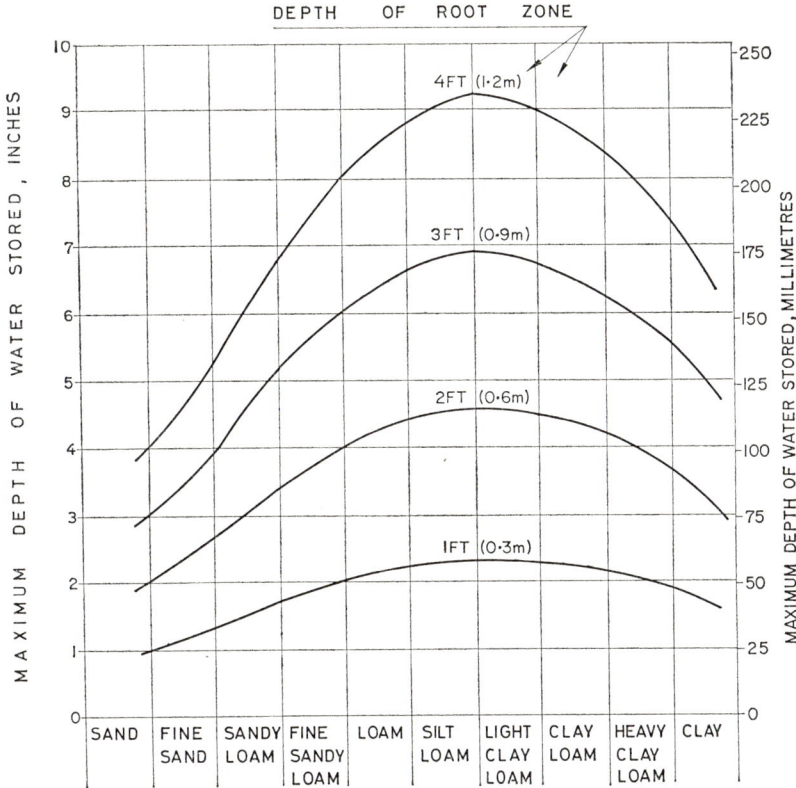

The depths of water available to crop between soil field capacity and wilting point have been based on data from U.S. Dept. Agr. 1955 Agr. Yearbook p.120.

Fig. 47 Available water storage capacity in relation to soil texture and depth of crop root zone.

characteristic irrigation requirement, which may be termed the 'ideal requirement', there is inefficiency, loss of water or crop growth and detriment to the economy of the project.

The application diversity factor may, therefore, be represented as:

$$F_{ad} = \frac{\text{Ideal basic irrigation requirements} - \text{Excess or deficit in individual actual irrigation deliveries to crop-land unit (field farm or group)}}{\text{Ideal basic irrigation requirements}}$$

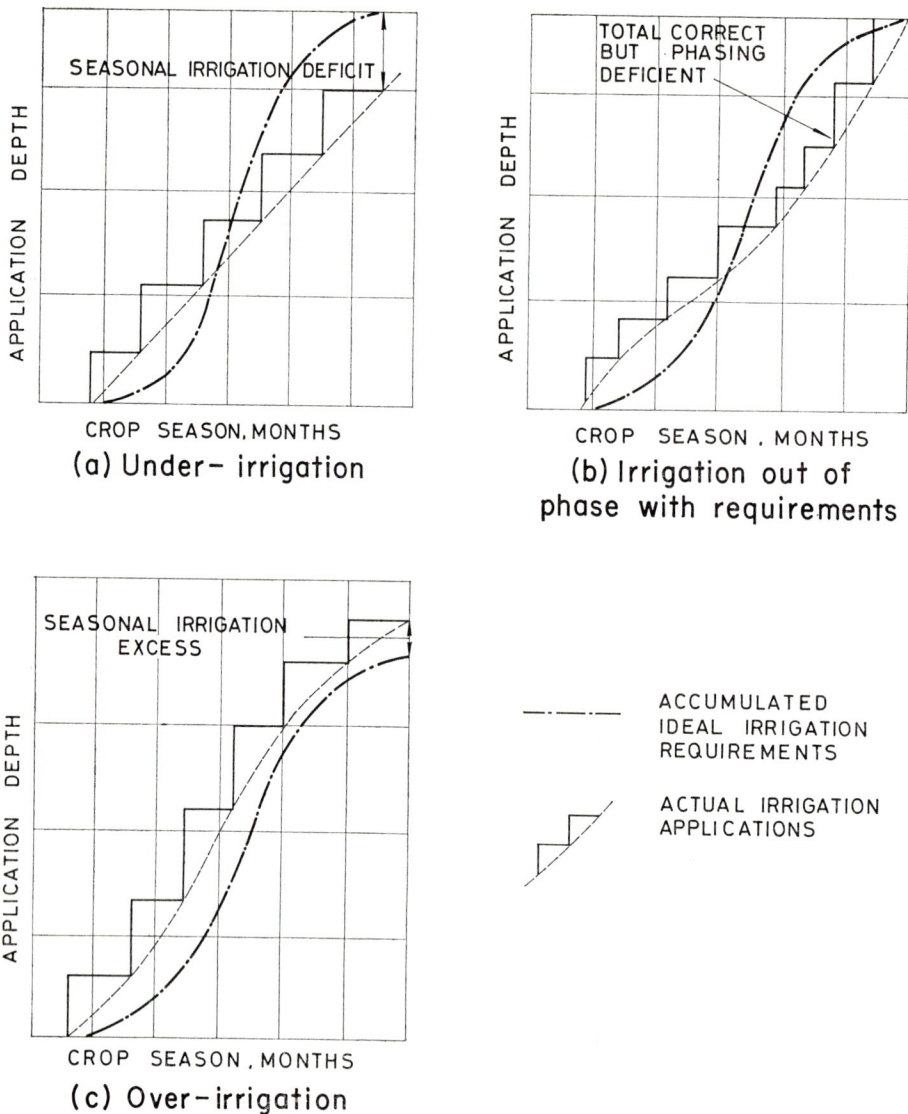

Fig. 48 Diagrammatic illustration of irrigation demand/supply phasing problems.

Apart from taking care of diversities in actual practice, such a factor will serve to take into account the following:

(i) Loss of crop yield and increase in salinity hazard due to gross seasonal under-irrigation.

(ii) Loss of crop yield and waste of water due to incorrect phasing of irrigation as regards timing and amount (see Fig. 48).

(b) *A soil storage factor*

There appears to be a case for isolating a factor to take account of the following losses which influence the application efficiency:

   (i) Seepage and wastage from watercourses,
  (ii) Run-off from fields,
 (iii) Deep percolation below the root zone,
 (iv) Evaporation from the soil surface,
  (v) Interception and dispersal by wind in the case of sprinkler irrigation,
 (vi) Special applications for purposes of leaching out salts.

The soil storage factor may be expressed as follows:

$$F_s = \frac{\text{Volume of irrigation supplies retained in the root zone and available for use by the crops}}{\text{Volume of irrigation deliveries to the field unit (field, farm or group)}}$$

(c) *A uniformity factor*

In addition, there are the following influencing factors, arising from cultural practices, of which account should be taken when assessing overall application efficiencies:

  (i) Uneven distribution of surface water applications resulting from inadequate grading of field surfaces and causing local under-irrigation and corresponding loss of benefit.
 (ii) Lack of uniformity in water application (and relatively high surface-evaporation losses) where frequent shallow irrigations are necessary due to shallow rooting depth of crops (particularly if soil storage capacity is low also), or due to limitations of the distribution system.
(iii) Adverse effects of prevalent high winds on the distribution pattern where sprinklers are used.

The uniformity factor may be represented by the following expression:

$$F_u = \frac{\text{Volume of the root zone effectively moistened by irrigation}}{\text{Volume of the root zone of the crop}}$$

Revised application efficiency

Combining these factors, the revised composite expression for application efficiency becomes:

$$E_a = F_{ad} \times F_s \times F_u \times 100 \text{ (per cent)}$$

The range of values which can be expected generally for the above three factors (and discounting the human tendencies of irrigators to 'hoard' water by over-ordering so as to be on the safe side) is of the following order:

F_{ad} = Application Diversity Factor = 0·60 to 0·95
F_s = Soil Storage Factor = 0·40 to 0·85
F_u = Uniformity Factor = 0·70 to 0·95

and the resulting values derived for application efficiencies, E_a, will generally lie in the range 20 to 75%.

CONVEYANCE EFFICIENCY

The degree of efficiency attained on any conveyance system will depend in the main on:

(a) The proportion of water lost by escapage through wasteways and over spillways;
(b) The extent of seepage and evaporation opportunity which exists in the conveyance and distribution systems;
(c) The extent to which actual irrigation deliveries throughout the project exceed or fall short of the ideal irrigation water requirements for maximum agricultural production. In this connection, the efficiency of the conveyance system is necessarily bound up with and affected by the efficiency and phasings of field applications.

The economic aspects of conveyance efficiency become increasingly important in feasibility considerations, particularly in multipurpose projects where alternative uses of water may offer greater economic benefit. If the proportion of water lost in a conveyance and distribution system amounts to one-third, it follows that the quantity delivered could be increased by 50% were all such losses to be eliminated. In the event of revenue being dependent on quantities delivered, the consequent increase in return on capital could be significant.

Unrecoverable conveyance losses due to seepage from canals and the effects of water weeds, apart from reducing both the irrigation efficiency of the project and the overall production of the project area, constitute a waste and abuse of reservoir and canal capacity. It is worth bearing in mind that a continuous loss of only 39 l/s (1·38 cusec) due to such causes represents a wastage of 1·2 million cubic metres (1000 acre-feet) a year.

As there is a difference in characteristics of losses, the conveyance systems are usually considered under two distinct sections:

(i) Major conveyance structures.
(ii) Distribution system.

Conveyance system

The means of transferring supplies from source to the distribution system have a direct bearing on the conveyance efficiency obtainable. In the case of a project where releases from an impounding reservoir are discharged into the river and then diverted into an unlined canal at a barrage some distance downstream, significant losses can occur from seepage, evapo-transpiration from water weeds, and from evaporation.

The conveyance efficiency is also certain to be lowered as a result of seepage from lengthy unlined main canals when these are located in permeable soils, and especially where the water table is well below the canal bed level.

In particular, the problems of controlling canal level and discharges, in relation to demand, will be increased in proportion to the canal length and the complexity of the distribution system. The escapage losses which result from overall operating deficiencies in phasing of requirements and supplies will detract from the conveyance efficiency of the system. There appears, therefore, to be a case, as in the instance of application efficiencies, for introducing a diversity factor to draw attention to, and ensure improvements in, phasings of supply and demand.

Distribution system

Water losses in project distribution systems are in general due to causes similar to those affecting the major conveyance structures, but there is frequently a difference in degree.

The proportion of losses tends to be higher because:

(i) The proportion of escapage tends to be higher, especially if water is supplied on a demand system, or if the area is subject to local storms during the irrigation season;
(ii) The canal sizes are smaller and wetted areas are larger in relation to the unit discharge, giving rise to greater seepage opportunity.

Conveyance Diversity Factor

A revised expression for overall conveyance efficiency is suggested below:

$$E_c = \frac{W_d}{W_r} \times F_{cd} \times 100 \text{ (per cent)}$$

where

F_{cd} = Conveyance Diversity Factor (similar to Application Diversity Factor)

$$= \frac{\text{Ideal irrigation requirements calculated at source} - \text{Accumulated excess or deficit in actual releases at source}}{\text{Ideal irrigation requirements calculated at source}}$$

The range of values which can generally be expected for the above factors are:

$\dfrac{W_d}{W_r}$: 0·60 to 1·00

F_{cd}: 0·70 to 0·95

The resulting values derived for the conveyance efficiency (E_c) will generally lie within the range 40 to 95%.

The order of application efficiency (E_a) has been previously given as 20 to 75%. The irrigation efficiency (E_i) is, therefore, generally expected to be within the range 10 to 70%.

SCALE OF BENEFITS TO BE DERIVED FROM IMPROVED EFFICIENCIES

A practical illustration of the potential scale of benefits envisaged as arising from improved irrigation efficiencies is well-demonstrated by the findings of the 1960 Select Committee report to the U.S. Senate[8] which gives the following projections to the year 2000 compared with existing conditions in 1954.

Table 31
PROJECTED IRRIGATION WATER REQUIREMENTS IN THE
U.S.A.; 1954 AND 2000
(also see Table 17(a) in the Appendix)

Description	1954	2000
Irrigated hectares	11 980 000	22 380 000
Storage and diversion (m³)	240 000 000 000	245 000 000 000
Irrigation (m³)	141 200 000 000	169 200 000 000
Application efficiencies		
Eastern Region (%)	60	70
Western Region (%)	45	60

Thus, in spite of almost doubling the 1954 area, it is estimated that relatively insignificant increases in stored and applied irrigation water will be required. This will be achieved as a result of raising applied efficiencies from 45/60% in 1954 to 60/70% in 2000 through improved techniques, technical assistance and general education.

It is an interesting exercise to apply these principles broadly to a country like West Pakistan where, apart from a desperate water shortage, a severe limiting factor to progress is the scarcity of suitable dam and reservoir sites on her three Eastern rivers. If one assumes a present application efficiency of 50%, and an irrigated area of 81 000 km² (20 million acres) requiring annual diversion of 123×10^9 m³ (100 million acre-feet) for irrigation then, by raising the application efficiency to 60%, the additional water so made available would be of the order of 21×10^9 m³ (17 million acre-feet), or nearly four times the live storage capacity of the reservoir recently completed on the Jhelum river at Mangla. Additional benefits which would accrue are improved crop yields per hectare and hence increased total agricultural production.

SOME PRACTICAL ASPECTS CONCERNING IMPROVEMENT OF IRRIGATION EFFICIENCIES

Arising from the above considerations of efficiency factors, and bearing in mind the ever-increasing need for maximum benefit per unit of water released or diverted, it is pertinent to make some comments about sundry practical aspects relating to improvement in irrigation efficiencies.

(a) Application efficiencies

Reduction of farm distribution losses

The loss of water by seepage from farm watercourses can be expected to be proportionately more than from distributaries, because the wetted area per unit volume of flow is generally greater than in larger canals. The seepage opportunity is, therefore, correspondingly greater. Escapage losses from farms can often be attributed to excessive rate of delivery in relation to the rate of application required at the fields. Water savings resulting from lining and reduction of escapage would:

(i) Decrease water costs per unit of crop yield;
(ii) Tend to reduce water demand from farms; or
(iii) Increase the irrigated area where land is available.

The operational requirements of lining to prevent seepage from farm watercourses are not so stringent as for larger canals, and the range of suitable methods is, therefore, wider. As an alternative to cheap compacted earth lining, the use of cement or lime stabilized soil, if properly constructed, has the advantages of low cost and reasonable durability. The use of lime as a soil stabilizing agent in this connection has not perhaps received as much attention as it deserves, having regard to the fact that it has been used so effectively in road bases in the U.S.A. and in Southern Africa.

Greater flexibility of delivery head would help to reduce escapage losses, although this may entail the use of adjustable water metering devices or variable modules at farm turnouts.

Reduction of field application losses

Significant reductions of application losses can be achieved by cultivators as a result of a wider appreciation of the physical characteristics of the soils under their control. Field run-off can often be minimized or eliminated by reducing the application rate to match the infiltration capacity of the soil, which in itself may sometimes be improved by cultural practices. In the case of heavy clay soils with low infiltration capacity, the effective slope and dimensions of the fields should be arranged to allow for the corresponding tendency towards high run-off. Where some run-off is inevitable, effective use can still be made of the excess water by collection and recirculating to the field by pumping, impounding on waste or saline land for fish culture which can be a valuable source of protein, by application to a 'tailwater' rice crop, or by disposal to a groundwater spreading area where a low water table permits this.

Excessive percolation beyond the root zone should be largely avoidable by matching the depth of each application with the available water storage capacity of the soil in the root zone of the crop at the time of irrigation. This requires some observation and knowledge of the moisture-holding capacity of the soil, which has been referred to previously in connection with the estimation of application rates. Persistent over-irrigation not only gives rise to low water application efficiency, but can also cause depression of yields as a result of leaching of soil nutrients and, if the practice is widespread, may result in waterlogging.

Losses due to evaporation from the soil surface are not easy to avoid or minimize. During irrigation it is evident that a soil with a high infiltration capacity will give rise to less evaporation opportunity from the free water and soil surface than a tight impervious soil. Some advantage is, therefore, likely to be gained in this direction by increasing the infiltration capacity of heavy soils by cultural practices designed to increase soil aggregation. Evaporation losses from the soil surface after irrigation are also reduced by crop trash or other mulching. This forms a moist, micro-climate barrier between the soil and the atmosphere, and is particularly effective with young crops when the proportion of bare soil is high in relation to cover. For some high value horticultural crops even plastic film is being used to act as a mulching cover.

The rate of evaporation loss from the surface varies throughout the day and season, depending upon the humidity and temperature. The proportion of evaporation loss to the depth of application will be higher during daylight irrigation under clear skies in the hottest part of the season than at night or under cloud.

With sprinkler irrigation, excessive losses can occur through leaf interception, evaporation and dispersal by wind. In conditions of continuous high wind in daytime, these losses can be very detrimental to an otherwise efficient method of water application. Their effects may be overcome to some extent by the design of the layout with regard to the spacing and discharge of sprinklers, and by careful attention to the aspects of the droplet size and radius of throw.

The necessity to ensure freedom from the effects of accumulated salts in the root zone requires leaching application to be made on some projects. To avoid wastage of water and the cost of surplus capacity, consideration should be given to timing the leaching operations so that 'off peak' water is used in conjunction with rainfall if possible.

Timing and depth of application
High water application efficiencies and good yields, even on the best of soils, cannot be obtained unless the irrigation requirements of the crops are continually provided for, without waste, from the available moisture in the crop root zone. The rate of crop evapotranspiration, which varies throughout the growing season, must be matched by the rate of application of irrigation water. The permissible interval between irrigations and the required depth of application are dependent on the soil texture and the crop rooting depth, as well as on the rate of evapo-transpiration.

It is considered that proper timing of applications and a reasonable estimate of the minimum and mean depths of water to be applied to the fields are two of the most essential prerequisites to the achievement of high application efficiencies and yields. This matter is also referred to in Chapters 2, 3 and 8.

In practice, when dealing with efficiency factors under operating conditions, it is necessary to make a clear distinction between design (generally engineering) characteristic curves of ideal requirements and those used for a particular period of individual year for purposes of operation (agricultural). Since calls for water from source have to be made in advance of accurate knowledge of climatological conditions, it is also necessary to consider practical ways and means of making periodic accountancy checks to correct deviations from the ideal.

For project planning (design) purposes, basic irrigation requirements are usually estimated conservatively on dry year conditions.

Under actual operating conditions, crop water demand for the particular period will be related to the prevailing climatological conditions, which will not necessarily be related in any manner to, say, average or other characteristics determined from analysis of past statistics. As a first step, then, it becomes necessary, when calling for water releases, to make or interpret as reliably as possible a set of weather forecasts for the accountancy period envisaged. The balancing of the soil moisture account becomes a matter of countering deficits in evapo-transpiration (consumptive use) with credits from rainfall and irrigation. Running logs of soil moisture and climatological conditions should be kept scrupulously throughout the growing periods. Adjustments to calls for future releases should be made at the end of each accountancy period, which preferably should be as short as possible consistent with the overriding requirement that growth of the crop should be uninterrupted. The soil moisture in the root zone must be maintained between field capacity and the wilting point.

To a certain extent, the water in the root zone acts as a balancing reservoir, and therefore, applications may be varied to a minor degree as regards depth and timing to suit water delivery schedules, provided that the accumulated balance is maintained between the crop/soil requirements and the water supplies.

One of the most difficult judgements to make in these balance-sheet adjustments under present circumstances is the amount of rainfall that is really 'effective', *i.e.*, which becomes available in the root zone. The subject presents a field of research, the results of which could be of great practical value to designers and cultivators.

A typical practical application of the adjustment procedure is illustrated in Fig. 49. The soil moisture condition in the crop root zone is plotted against time for the accountancy period.

Uniformity of application
Undulations in the soil surface micro-relief, and consequent lack of uniform distribution of water on the field, is a common cause of reduced application efficiency and uneven crop yields under surface irrigation. This defect does not arise to the same extent where sprinkler irrigation is used, and it can be largely remedied by grading the surface to

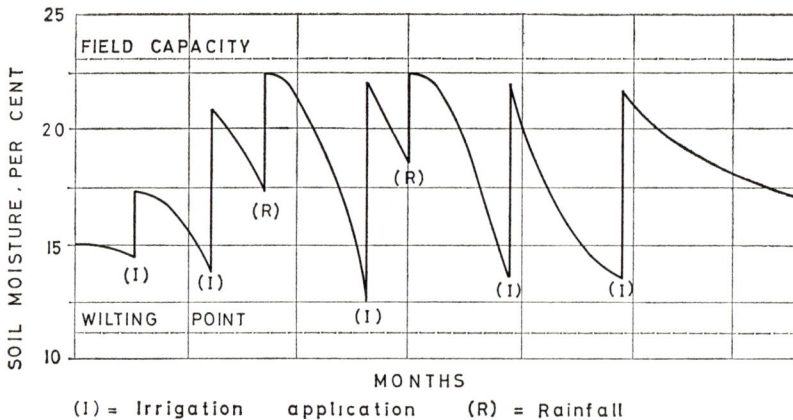

Fig. 49 Soil moisture variation diagram.

remove high and low spots. Further improvement can be effected by the use of large modern land planes to reshape the whole field surface to uniform grade at optimum slope, subject to the topsoil depth being adequate.

Efficient, uniform distribution is difficult to achieve with surface application on soils of high to medium infiltration capacity unless a high rate of discharge per unit area can be made on to the field, and rapid application can be effected.

In the interests of efficiency, the rate of application per unit of area should be directly proportional to the soil infiltration capacity, and should be limited only by considerations of sheet erosion of the field, and adequate control of the stream.

Field, basin and border check dimensions should be restricted and designed so that the shallowest practicable depth can be applied with the minimum anticipated rate of delivery flow, since it will otherwise be difficult or impossible to make small applications without serious lack of uniformity. The sandier the texture, the greater the diversity becomes, particularly where very low slopes are involved, due to the familiar 'wedging' effect resulting from the time taken by the water to travel the length of the field to be covered.

Except in the case of application by sprinkler or on clay soils, shallow waterings, therefore, generally lead to low application efficiency. For this reason, deeper waterings are to be preferred, provided that they do not exceed the depth of the root zone (except for the purpose of salt leaching). For the same reason, it is advantageous for irrigated crops to be fairly deep-rooted, and to have as extensive and dense a root system as possible.

One suggestion for a simple method of effecting more rapid and uniform applications of water on permeable soils is illustrated in Fig. 50. Shallow furrows are constructed down the slope by means of a tractor-drawn roller designed for the purpose. The furrows are tie-ridged at the upper end and midway down the slope. The water will travel down

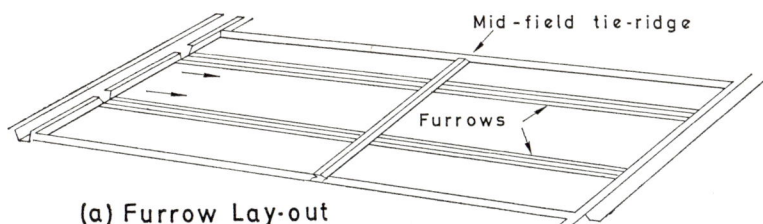

(a) Furrow Lay-out

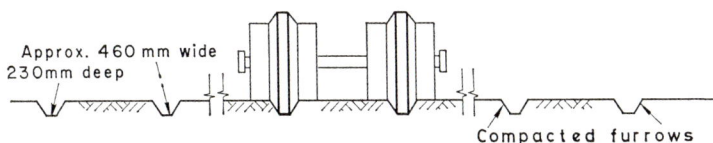

(b) Furrow Roller

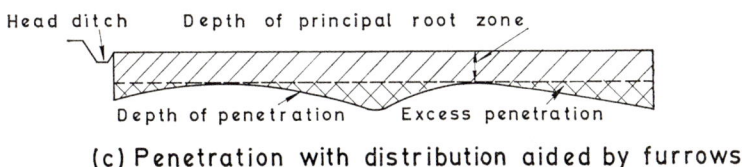

(c) Penetration with distribution aided by furrows

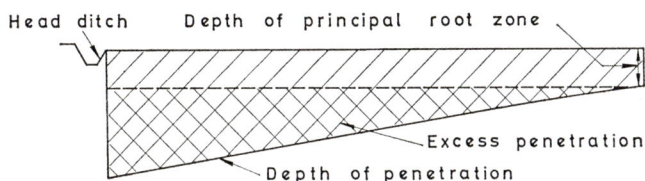

(d) Penetration with normal sheet flow

Fig. 50 Proposal for use of furrow rollers to improve uniformity of distribution on permeable soils.

the field in the furrows more rapidly than by sheet flow, and supplies will reach the inter-mediate tie-ridge, concentrate, burst the tie-ridge and continue down the furrows to the end of the border or field. Concentration will then tend to pond-back up the slope, creating the distribution pattern shown in Fig. 50(c). The improvement over the usual distribution is indicated by comparison with Fig. 50(d).

(b) Conveyance losses

The methods which can be adopted to reduce conveyance losses are basically: canal lining, water weed eradication, reduction of escapage by the use of a balancing reservoir, and improved gate control.

Canal linings

The development of lower-cost canal linings will continue to be a valuable service in the interests of more efficient irrigation. Although compacted earth linings are among the simplest and cheapest to construct where suitable material is available, they are reported to be generally quite satisfactory in use. They have the disadvantage, however, that over-excavation of the canal section is generally necessary. The prevention of damage to the cheaper forms of lining due to external water pressure when the canal is empty, and from animals such as water buffalo in Pakistan, is also a problem which has not yet been entirely solved.

The production of a cheap and permanently satisfactory canal sealant for application while the canal is in service would also be a valuable development. Some means would preferably have to be found to enable the chemical to penetrate about 0·6 m (2·0 ft.) below the surface before solidifying, thus providing sufficient cover to prevent damage.

The use of cationic bituminous emulsion seems to have possibilities in this field, especially when applied in conjunction with electrodes temporarily inserted below the bed of the canal. The flow of electric current causes the emulsion to penetrate well below the surface, and thus improves the durability of the seal.

Water weeds

The losses due to water weeds will obviously depend on the area of vegetation involved, and where this is significant, it may be profitable to eradicate them either by chemical or by mechanical means.

Balancing reservoir

Where the topography is suitable, the provision of a balancing reservoir in the conveyance system may contribute to the improvement of phasing of demands and supplies. A minor reservoir of this type, when located on the main canal and close to the distribution area, acts in much the same way as a surge chamber in a hydro-electric development. Short-term fluctuations in distribution demand can be met or absorbed by the reservoir without having to make similar variations in the flow of the main canal. Spillage resulting from cancellation of demands can, therefore, be reduced, and sudden increases in irrigation requirements can be met until the main canal discharge is adjusted. This form of regulation is also valuable where a small hydro-electric station is located at a canal drop.

Gate control

The standard of water discharge control achieved in the canal system is certain to be reflected in the irrigation efficiency of the project. Good control of the water level in canals is necessary for economical water-use and consistent supply of the requisite quantities to distributaries and farms. Gated regulators which are automatic in operation, and which control either the downstream or upstream water level, have been successfully marketed for some years. For large canals, it is to be expected that electronic controls will provide more rapid and efficient methods of interlinked operation of gates, which should contribute materially to the reduction of escapage losses.

(c) Peak supplies

In the case of projects on which a pronounced peak demand is experienced for two or three months in the year, there may be difficulty in providing entirely adequate supplies, since a disproportionately large canal capacity has to be provided for a short period of use. This period is, however, usually critical for crop yields. In these circumstances, it may be advantageous, from the aspects of cost and efficient use of water, to provide peak supplies by means of tubewells discharging directly into the distribution system. This matter is also referred to in Chapter 6.

This method can be adopted where the ground water is at an adequate depth from the surface and is of suitable quality, and when the specific yield of the aquifer is satisfactory for pumping.

The annual cost of providing the additional canal capacity per unit volume of peak supply may be greater than the cost of providing the same quantity from tubewells. This aspect is illustrated diagrammatically in Fig. 51. Incremental annual costs per unit of volume supplied (obtained from the incremental capital costs A and A + B) for constructing a canal of the required capacity to supply the corresponding incremental demands, are compared with the equivalent cost of tubewell water. It would appear that this form of development, where potentially feasible, is likely to be most advantageous in the case of fairly long canals of medium capacity, that is up to say 198 m³/s (7000 cusec) and of the order of 80 km or more in length.

(d) Water charges

There is a growing trend of opinion that the scale of water charges is likely to play a dominant part in the improvement of both application and conveyance efficiencies. It is generally in the common interest that cultivators should pay for water according to the quantity delivered and not, for instance, by means of a cess on crops produced or on the area irrigated. It is also desirable that water charges should be as high as practicable and should be closely related to the actual cost of supply. These considerations are paramount in the encouragement of improved efficiency. Another inducement towards more economical use of water is to apply a scale of rising charges for water supplied if the quantity delivered exceeds a stipulated total based on area irrigated.

(e) Other factors

It should be recognized that the efficient application of irrigation water is only one factor amongst several which are necessary for the achievement of high crop yields. The expenditure of effort and money on engineering and administrative improvements as regards water application must be accompanied by other equally important improvements directly concerned with water use by crops. These include cultural practices such as pest and weed control, preplanting cultivation and the application of fertilizers, aimed at increasing the depth and nutrient status of the root zone, and thus increasing crop yields

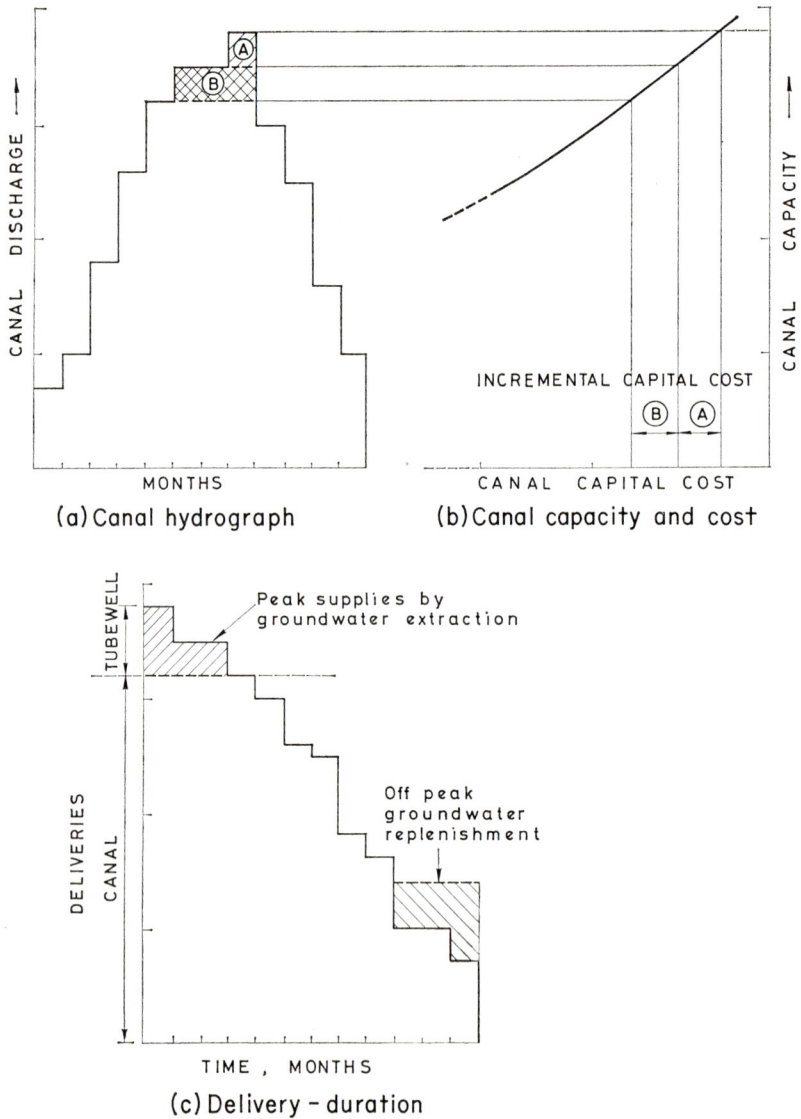

(a) Canal hydrograph

(b) Canal capacity and cost

(c) Delivery – duration

Fig. 51 The provision of peak irrigation supplies.

and effective soil moisture storage. Advances in plant breeding directed towards improved yields and rooting habits under irrigated environmental conditions could also contribute greatly in this respect.

OUTLOOK

There is considerable scope for increasing agricultural production by effecting general improvements in irrigation efficiencies. By this means, productivity per unit of water may be raised and the savings in existing water supplies will become available for use on extensions to present developments or for new projects. Apart from this consideration, the unit value of water at source will also be increased. This is considered to be essential in order to improve the position with respect to securing allocation of storage capacity and diversions for agricultural purposes, particularly where there is competition for available resources from other interests.

In order to achieve higher irrigation efficiencies, it will be necessary to concentrate on methods designed to improve the phasing of supplies with respect to climatological, crop and soil characteristics, and to step up the campaign at all levels for reduction of avoidable water losses. If the high standards envisaged are to be achieved, this will mean intensive programmes of technical education and encouragement at farm levels, and the fullest co-operation of many professions.

8 Research in the fields of water resources engineering

THE NEED FOR RESEARCH

The survival of an industrial programme, whether it is a chemical industry or the manufacture of hard goods, depends on the most efficient and effective use of every unit of raw materials that goes into their product. Industry must find lower cost raw materials which require minimum modification and processing to sustain competitive survival. No industrial firm can compete unless it has a reliable produce which is worth just a little more, will last just a little longer, and with a little less repair and maintenance than a competitor's produce. In essence, research in industry is a key to progress.

Similarly, in irrigation and drainage improvements in arrangements for collection, diversion, distribution and application of water; improvements in fabrication and assembly of irrigation and drainage equipment; improvements in forecasting demands and accounting for deficiencies in correct rhythm with climatic and storage time disciplines parallel the normal industrial approach to meeting competition.

Generally, it appears that water research in all its branches, and especially basic research, is being dangerously neglected. If one looks through reports of symposiums held at various times during the fifty years 1910–60 it becomes apparent that the same fundamental questions, relating to research outlook, recur with monotonous regularity. Thus irrigation and drainage research is still concerned with the same basic problems of half a century ago.

In these circumstances it seems evident that creative research will be necessary to improve the stature of irrigation and drainage research.

Creative ideas and solutions transcend the established order of things and therefore make it necessary to proceed along new and unconventional lines. It requires challenging of the existing order of things and of established routines.

As long ago as 1910 Sarn Foster wrote:

Thousands of acres can never be reclaimed until the duty of water upon lands already under irrigation has been raised by reducing the losses due to seepage and evaporation in bringing the water to the land and applying it by devising and introducing better and more economical methods of distribution and application, and by preventing the wasting of water so commonly practised by irrigation in order to hold the entire amount granted by state statutes passed when water was plentiful, settlers few, and methods crude. Even today, thousands of acres are being waterlogged and alkalied, and crop yields reduced by the use of too much water; while adjoining lands of equal, or even greater, fertility can be used only for range, owing to the lack of water.

Any references to research elsewhere have been concerned mainly with highlighting some specific trends or achievements to date. It is, however, considered necessary to concentrate on the specific subject and to peer into the future with the object of trying to discern fields of endeavour in seeking steady progress as regards improved scientific and economic efficiency in the utilization of resources.

THE APPROACH TO RESEARCH

First of all, there must be pure, or basic, research—which is research not directed at a specific, preselected goal, but rather aimed at discovering new facts about the laws of nature and their operation.

Secondly, we must have a co-ordinated programme of applied research, the objective being to take the laboratory findings of the basic research people and turn them to practical use. And finally, we should follow up with a broad and intensive programme of putting the newly developed techniques and procedures to work. Abstract knowledge of what to do accomplishes little good if it is not used.

There is the usual question: Who should undertake basic and who applied research? Generally the latter has been sponsored by industry and the former used to be the prerogative of the universities. The pattern seems to be changing with governments in one form or another taking an increasingly large interest to ensure that the minimal amount of basic research is undertaken. Discoveries from applied research are not always disseminated and even if disseminated are not always of universal interest. Basic research is a fundamental requirement in most countries today to ensure best use of national resources.

From the outset it must be clearly understood that it is no longer feasible to segregate research for purely irrigation and for industrial uses. Furthermore, in any approach to research connected with water resources engineering, the emphasis can no longer be mainly or exclusively on 'use' of water. Attention must be focused with increasing attention on 're-use' and hence 'rehabilitation' of water. The comment from some regions to the effect that, in order to be able to re-use water, there must be some to use in the first place, is valid.

In considerations of re-use of water it must be borne in mind that industrial developments have been responsible for deterioration in quality of water inherited downstream for agriculture. This has heightened competition for 'first' use of limited resources.

In countries where water has become a limiting factor to future development the consumptive use criteria would not appear to be appropriate, with particular reference to industrial use. To reserve large quantities of water for the purposes of diluting undesirable wastes so as to maintain an adequate standard of water in the rivers could bring development to a halt in certain areas. It is therefore suggested that more emphasis be placed on 'water rehabilitation', the principles of which are shown diagrammatically in Fig. 52. The figures used have been chosen at random to illustrate the principle and do not represent the actual situation at any specific location. It will be seen that, assuming

F

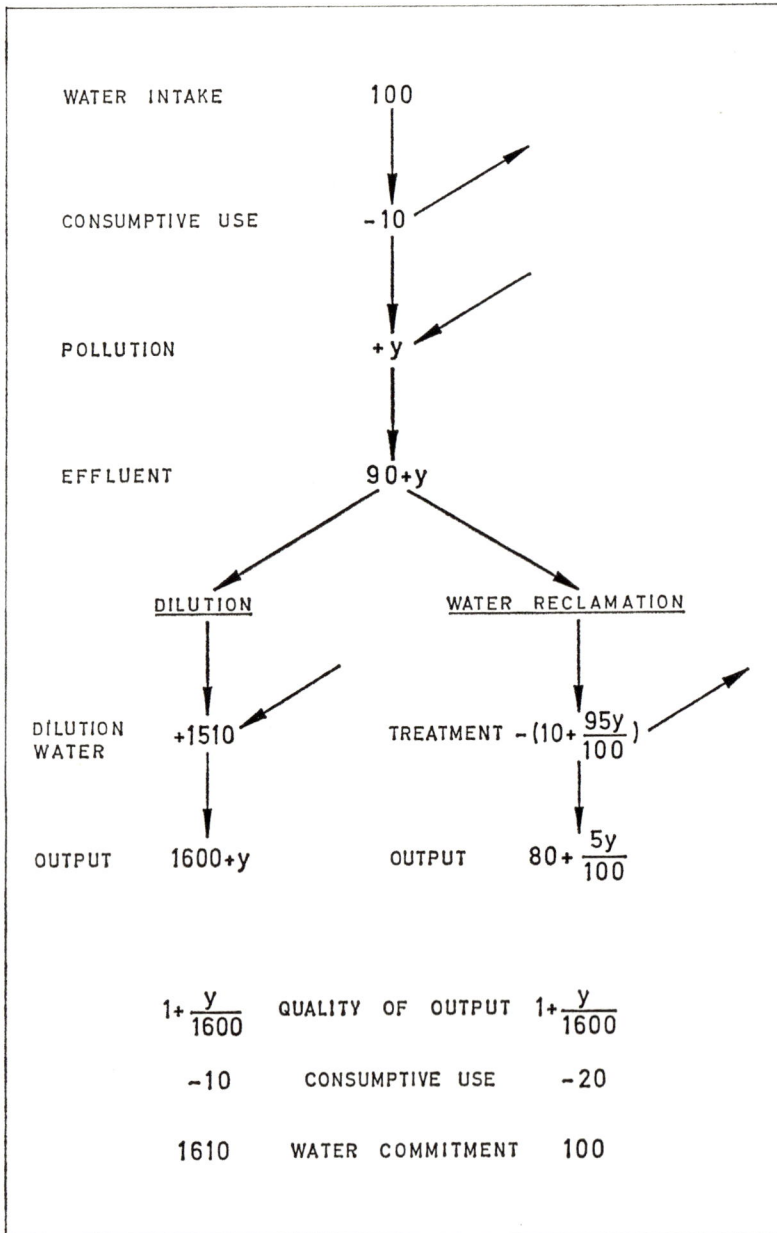

WATER INTAKE   100

CONSUMPTIVE USE   −10

POLLUTION   +y

EFFLUENT   90+y

DILUTION   WATER RECLAMATION

DILUTION WATER +1510   TREATMENT $-(10+\frac{95y}{100})$

OUTPUT 1600+y   OUTPUT $80+\frac{5y}{100}$

$1+\dfrac{y}{1600}$	QUALITY OF OUTPUT	$1+\dfrac{y}{1600}$
−10	CONSUMPTIVE USE	−20
1610	WATER COMMITMENT	100

Fig. 52 The principles of water rehabilitation.

identical use, the identical quality of final return water, the application of the principles of rehabilitation will permit 16 times more activity (development) than in the case of dilution, even though the consumptive use of the process may be twice as great.

At the present time the cost of rehabilitation may be greater than that of dilution but in the long term when the yield of new water per unit of storage declines as the storage capacity provided or available approaches the point of marginal return the position will be reversed; and the rehabilitation process will raise the ceiling of permissible activity which is dependent on a given source of supply.

Increasing the crops

The function of irrigation is to improve crop quality and quantity per unit of water per unit of land.

The greater part of the irrigated areas of the world was developed for cropping intensities that are lower than the land is capable of sustaining and also lower than can be tolerated in the face of the population and nutritional pressures foreseen. Generally, the irrigated areas are served by complex systems of canals designed to make use of the natural flows of the rivers. Now, by the addition of storage reservoirs and water table control, together with associated advances in agriculture, crop production from these areas is being intensified.

Except in the least developed territories, large-scale irrigation cannot be separated from overall water resources planning. Economic (as distinct from purely financial) evaluation of storage, distribution and infrastructure projects in relation to resultant benefits automatically involves public funds and therefore, national considerations. This has led, first, to multipurpose engineering, and recently to the concept of integrated planning of river basin developments, to identify and formulate those projects with the highest priority for the allocation of scarce resources: water, finance and implementation capacity.

Use of computers

Developments of computers have made it possible to undertake within a reasonable time the massive volume of calculations required to make a proper analysis of complex irrigation and multipurpose systems, taking account of all foreseeable constraints.

A large number of trial solutions with different parameters can be examined in a very short space of time as was the case in the 1966 World Bank Study of Water Resources of West Pakistan.

SOME AVENUES FOR RESEARCH

It is not feasible to cover comprehensively all possible fields of research which, by definition, must generally remain open ended to cover the widest spectrum of possibilities.

The following examples are selected at random merely as an indication of the general scope and some of the lines that should be pursued in the light of present engineering experience.

Soil intake rate and the use of root zone storage

If we could just measure, predict, and control accurately the water intake characteristics of soil, irrigation efficiency would be improved immensely.

In classic research on intake there have been many approaches about which we have been hearing for years. But have we really explored the avenues along which the chemists and physicists have proceeded? For example, what are the forces that really influence the movement of one molecule of water in close relation to a soil particle? What are the physical forces which must be isolated and measured before we thoroughly understand the nature of water movement into the soil?

There remains the practical problem of maintaining a most suitable water and air environment in the soils subject to compaction caused by machinery and water application. Soil fertility needs have been determined for the major elements, but it is possible that the minor elements may raise production to another plateau if soil and water can be properly managed.

From a different angle it will also be deemed worth while to study the long-term effects on run-off and movement of underground water of soil compaction by heavy farm machinery and the massive use of chemical fertilizers.

We know now the time disciplines of basic water requirements for any climatic complex and for a definite locality defined by latitude. What about effective rainfall, micro-climatic effects and soil storage? How can we increase soil storage reservoirs and thus the time required for accountancy to permit correction of errors in indenting on precious storage and in assessing losses in transmission? Is it possible to develop deeper-rooted crops?

It has been noted that in a dry area like the Karoo of South Africa, lucerne, which was content for its roots to go to great depths in search of water, soon became spoilt when farmers started overwatering in respect of time disciplines, and the roots only went down 1 m to 1·2 m since it could with less effort get its water requirements from that amount of storage.

In another instance a bluegum tree on one side of a tarmac road sent a special echelon of roots under the tarmac road and up to within 1·2 m of a bed or roses on the other side which were being regularly watered. It must be remembered that the plant is a living organism and capable of adaptation to environment.

The United States Geological Survey has recently been able to fly a river and, through infra-red measurements and photographic recording, pick out where the colder ground-water enters that stream, thus mapping out the points of return-flow seepage quite accurately and effectively. Work by the Agricultural Research Service down in the Lower Rio Grande Valley indicates that infra-red photographic techniques also have considerable promise in detecting and mapping salinity-affected areas. Much groundwork is still needed to improve procedures of a breakthrough for practical achievement of soil moisture measurements over a wide area.

If we could make a photo-reconnaissance flight over any district in the area, say, on Monday morning at a certain hour, bring in the photos and run them through an electronic density scaler that has been precalibrated against field checks, a computer could then reduce the readout to a farm-to-farm or field-to-field basis. This will make it possible to prepare for the 'Water Master' of the district a chart showing how many cubic metres of irrigation water will be needed to satisfy moisture requirements at one end of this district on Tuesday and Wednesday, and the quantity needed in another area by Thursday. Such flights could be made as frequently as environmental conditions dictate. Similar techniques are already being used with computer programmes for the solution of the entire water basin planning studies. All that is required is to evaluate soil moisture status at a given moment in time over a broad area in order to integrate by automated methods the total field water requirements for any given period ahead.

Thus, after allowing for losses in reservoirs, canals and rivers it should be possible to give proper instructions to the gate operators to release in correct time order the proper quantities of water which have to travel a long way to the fields.

In order to ensure material improvements in irrigation practice it will be necessary to organize intensive demonstrations at farm level, by lectures and demonstrations of yields obtainable per unit of water on pilot farms which are so distributed within an irrigation complex as to be within easy access of the entire farming population.

As sophistication in practice develops based on a realization of the scale of benefits attainable through efficient distribution of moisture it will become possible to introduce automation of field supplies from furrows, feeders and main canals as a further requirement in the interests of elimination of water waste and reduction of labour requirements.

Underground water

About 60% of the US.A.'s total groundwater withdrawal is for irrigation. The total underground water resources within 0·8 km of the land surface are estimated to be about 100 times the total surface run-off. This underlines the need for strengthening research efforts to make this valuable source of water more readily available for multipurpose use. Even if the quality of such water is not perfect for a particular use, by applying it to an activity with less stringent specification it would release an equivalent of better water.

A major advancement in the study of groundwater has been the development of analog computer techniques to evaluate groundwater movement. Researchers can analyse the properties of an aquifer, its boundaries, its geometry, and the quantities and location of recharge and withdrawal. As an example, analogs are helping immeasurably in the complex studies of co-ordinating groundwater and surface water resources in the Upper Snake River basin of Central Idaho.

Much progress has been achieved as regards integration of steam and hydro-electric generation. Surely the same principles could be applied to surface and underground waters.

In the case of projects on which a pronounced peak demand is experienced for two or three months in the year, there may be difficulty in providing entirely adequate supplies, since a disproportionately large canal capacity has to be provided for a short period of use. In these circumstances, it may be advantageous, from the aspects of cost and efficient use of water, to provide peak supplies by means of tubewells discharging directly for crop yields. Such flexible arrangements for provision of peak supplies might be of considerable value.

This method can be adopted where the groundwater is at an adequate depth from the surface and is of suitable quality, and when the specific yield of the aquifer is satisfactory for pumping.

The annual cost of providing the additional canal capacity per unit volume of peak supply may be greater than the cost of providing the same quantity from tubewells. It would appear that this form of development, where potentially feasible, is likely to be most advantageous in the case of fairly long canals of medium capacity, that is up to say 200 m³/s (about 7000 cusec) and of the order of 80 km (50 miles) or more in length.

The question of recharge of aquifers needs to be studied with particular reference to the possibilities for diverting the wasteful flood peaks on some rivers on to suitable 'intake' areas. The factors affecting maintenance of a balance between depletion and replenishment need also to be studied. (See also Chapter 7.)

In the economic field little seems to have been done as regards establishment of proper criteria for the assessment of benefits and costs with respect to utilization of underground resources.

Waterlogging

Experience has shown that in flat terrain situated between large river tributaries a high water table is bound to result from indiscriminate field applications of water irrespective of time disciplines imposed by intake rate capacity or crop-soil requirements.

Figure 53 shows a comparison of the rate of rise of water tables in the Punjab region compared with growth of perennial irrigation.

As regards fields, the immediate essential is adequate drainage—not always an easy matter in flat terrain. Another more long-term solution is pumping by tubewells. This method has been successfully applied in the Salt River scheme of Arizona and is being introduced in the heavily waterlogged areas of West Pakistan.

'Wet-foot' crops (marsh plants such as papyrus and phreatophytes such as salt cedar arrowweed) are responsible for increasing evapo-transpiration, *i.e.* water loss considerably over that of normal 'dry-foot' crops.

Vast areas of the United States are inhabited by various species of phreatophytes which rob the country annually by millions of acre-feet of water and are therefore classified as pests from the water economy angle.

Experiments by Ebermayer in Germany and by Basov in Russia have shown the existence of cones of depression in the water under forests.

Mention has been made in Chapter 6 of the possibility of using the more salt tolerant

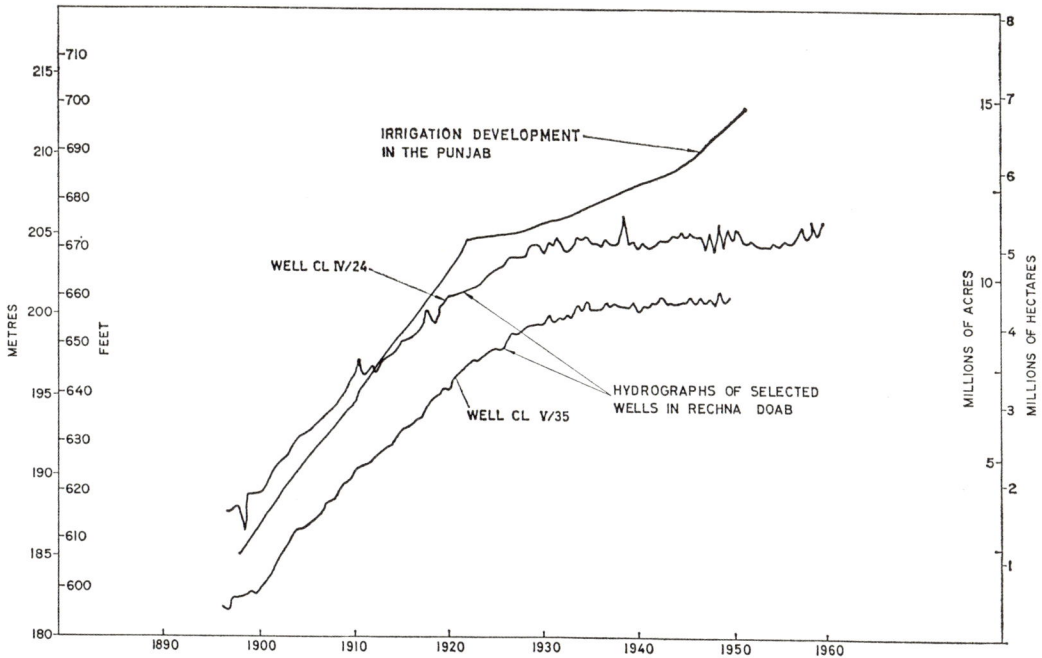

Fig. 53 Rate of rise of water tables in the Punjab, West Pakistan.

species of phreatophytes, planted along the edges of squares of cultivation in such a manner that the depression cone immediately under the perimeter plants causes a sufficient drawdown within the square to permit the cultivation of shallow rooted crops which are ordinarily sensitive to saline water.

Energy dissipation

Problems associated with dissipation of energy, whether of floods past dams or in canals and channels, have received attention through the years. These problems are particularly acute in the case of rock/earth dams founded on poor foundations.

It would seem necessary to improve our understanding of the mechanics of scour hole formation. It would appear that the depth of scour hole formed by a well distributed jet is not greatly affected by the type of foundation materials. But we do not know whether this holds for a mixture of soils with materially different resistances to scour. It would appear that the use of self-formed scour holes has great potential as regards energy dissipation but research work is needed to establish relationships to jet types to stabilized depths and shapes of scour holes and also the times required to stabilize such holes. This would lead naturally to investigations as to whether *preformed* scour holes could be successfully armoured using locally *in situ* or prefabricated materials.

We need to investigate whether continued heavy overspilling from great heights could lead to a form of fatigue in *in situ* rock whether in the bed or the walls of the valley.

Attention has been directed to the possibilities of 'distribution in place' of energy application. This involves a thorough study of the properties of hydraulic roughness. We are not clear as to what hydraulic roughness is and how to define it in terms of a surface composed of regular or irregular shaped rock or other objects packed in regular or random patterns. The problems of definition and measurement of hydraulic roughness have so far defied solution and the need for success in this field is becoming urgent.

Packing factor

During recent research carried out in England[57] which led to the development of a formula for the design of rockfill dams which can safely be overtopped, an important parameter, controlling the laws of flow with particular reference to dissipation of energy, was discovered. Three parameters were obvious: Rate of flow per unit length of crest (q_t); rock weight (d_s) and downstream slope (i). The fourth parameter was termed a packing factor, and is a dimensionless constant relating to the disposition in space of the rock.

The flow rate at which the stones began to move, *i.e.* the 'threshold' flow, q_t, was based on traction theory which takes into account the apparent weight of stones in water and their equivalent diameter, the angles of inclination and of repose of the stones, and a *packing factor*.

Thus the relationship between observed threshold flow, q_{ot}, and calculated values, q_{ct}, is governed by a dimensionless constant, P, which depends on shape and packing of the rocks, *i.e.* disposition in space. It was observed that a substantial change in q_{ot} could be achieved by manual rearrangement of the stones. This constant, defined as the packing factor, relates to the number of stones of a given size contained within a given area, P_c, or a given volume, P_{cv}. Thus:

$$P_c = \frac{\text{Unit area}}{\text{Number of stones per unit area} (=N) \times \text{area of average stone}}$$

$$= 4\pi/Nd_s^2 \text{ (for predominantly spherical shapes)}$$
$$= 1/Nd_s^2 \text{ (for predominantly cubical shapes)}$$
$$(d_s = \text{equivalent diameter of rock})$$

A particularly difficult challenge is presented with regard to dissipation of excess energy of water flowing in small channels. In these cases the Froude numbers generally lie in the range 2–4 which makes it difficult to achieve satisfactory results at reasonable costs. The energy loss obtainable from a hydraulic jump in this range is of the order of 10 to 20% of the total energy involved, compared with up to 80% in the case of large Froude numbers. Vertical stilling wells tend to fill with sand and debris. Flip buckets cannot be used because velocities are generally too low to cause the jet to spring clear.

Armouring

Under some circumstances small layers of graded coarse material may be used to stabilize the beds and banks of small channels. In rivers of moderate depth coarse material can be used successfully to stabilize the banks. Field observations have clearly demonstrated that a small amount of coarse material in a finer natural material may rapidly develop into an armour as scour and degradation occur. Available information on bed armour, as obtained in field and laboratory studies, should be documented in the technical literature. This natural phenomenon can in some situations be of great economic importance and the principles of armouring with coarse materials should be studied.

The use of armour for protection of 'soft' spots in river beds close to vulnerable structures or of scour holes may present great economic benefits.

Another related problem is prediction of retrogression or degradation. We are aware that short-term effects of an obstruction such as a weir across a silt satisfied river or channel leads to retrogression immediately downstream and accretion upstream of the obstruction. In the long term it has been observed in the Indian sub-continent and in the U.S.A. that accretion upstream of the 'arterial blockage' tends to flatten the slope of the channel but that in due course the channel tends to revert to its original slope but with a higher bed level. As the channel upstream reverts to its original slope more silt is released downstream causing a progressive recovery from retrogression. Generally the period of recovery is greater than that of retrogression, but, as in the case of the Khanki weir on the Chenab river, the eventual bed level downstream may be higher than it was originally. The problem is to predict the extent and timing of these developments.

Stabilization—fine sediment

The fine sediment transported by a stream is often referred to as washload. Generally only small quantities are found in the bed material and its concentration is essentially uniform throughout the flow cross-section. Studies thus far indicate that this material in the stream has significant stabilizing properties, the development of berms and the apparent viscosity of the water sediment mixture. The study of such and other properties would appear to merit further consideration.

With a concentration of the order of one hundred thousand parts per million by weight the apparent viscosity of the sediment-water mixture may be ten times as great as that of water alone. This large increase in 'effective' viscosity apparently reduces the fall velocity of the bed material making it much more susceptible to transport and consequently it reacts differently with the flow and is moulded into different bed forms than prevail for the usual case, perhaps affecting not only resistance to flow but also bed material discharge.

The ingenious use of killa-bush spurs for the stabilization of large canals in West Pakistan is an example of the formation of stable berms by induced fine silt deposit.

The head regulator for the Trimmu–Sidhnai Link of $310\,m^3/s$ (about 11 000 cusec) capacity was opened on March 3rd, 1965, and the discharge was slowly brought up towards full capacity. By March 16th, 1965, the discharge of the Link was brought up to approximately 140 m³/s (5000 cusec), but during this period high winds contributed to 'sloughing' of the canal banks.

On March 18th it was decided that the discharge of the Link would be reduced and measures taken to protect the banks.

Accordingly the flow in the canal was reduced to approximately 28 m³/s (1000 cusec) and killa-bush spurs were placed in the areas of the 'sloughing'.

The killa-bush spurs, consisting of double lines of poles, filled with brush, were constructed as required, perpendicular and parallel to the banks of the canal, at relatively close spacing.

They effectively protected the banks and permitted the forming of berms by silting between the spurs.

The construction of typical killa-bush spurs as on May 1st, 1965, is shown in Plate 5 and the effectiveness of berm formation by June 10th, 1965, is shown in Plate 6.

Economic criteria for viability studies

Great progress has been made recently in developing meaningful criteria for economic and financial analyses for rapid evaluation and ranking of projects. This matter has been dealt with more fully in Chapter 4. Nevertheless there is scope for further research particularly with respect to evaluation of such factors as nutritional standards, benefits and economics of providing recreational facilities and economic impact factors attributable to water resources projects.

PLANT, MATERIALS AND APPLIANCES

The greatest advances in dam engineering in recent times have been in the field of construction equipment for the mining, moving and placing of earth and rockfill where increased efficiency in handling large quantities of materials has resulted not only in holding the cost of rockfill and earth dams down to economic levels but also in making it possible to create large storage facilities on sites previously thought to be unsuitable for dam construction. Tarbela, the largest rockfill dam in the world, under construction on the Indus river in West Pakistan, requires a placing of three and a half times the rockfill involved in the High Aswan dam in Egypt in half the time.

In July 1963 Britain passed the Water Resources Act governing the abstraction of water from rivers and controlling effluent pollution. This has had a marked effect on water use practice and on the development of equipment. Farmers who were abstracting nearly 80 000 m³ (17 000 million gallons) for irrigation in 1965 and expected to double this amount in the next 35 years have, since then, developed an incentive to store their own water.

This has become practical and economic through new developments in equipment and materials such as polythene and butyl sheeting and tubing, refinements in sprinkler equipment, including portable systems, accurate metering equipment, plastic drainage pipes and a wide range of pumping equipment.

Polythene sheeting and butylite

Polythene sheet has many inherent advantages and also some limitations when compared with other materials that can be used as waterproof membranes in reservoirs. The two chief advantages are its very low cost per unit area and its availability in wide, seamless sheets. For water storage, the main properties are its inertness to chemicals, its retention of physical properties when buried and its resistance to attack by soil micro-organisms even in fertile soils. This means that polythene sheet can be buried indefinitely without deterioration. Once installed and buried it is unaffected by extremes of climate; severe frost will not embrittle it.

Polythene sheeting, although tough, can be punctured if not handled with appropriate care. This possibility can be overcome during its laying if the proper precautions are taken and subsequently, when installed, by covering the sheet with a layer of sand, gravel or soil.

A typical example of the use of polythene sheeting for reservoir construction is illustrated in Plates 7 and 8.

The advantages claimed for Butylite are:

High resistance to weathering, particularly ozone, ultra-violet degradation and oxidation.

Physical properties unaffected by temperatures from $-40°C$ to $+120°C$.

Successfully resists root penetration, even by the most aggressive forms of vegetation.

Unaffected by fungi, bacteria, soil acids, fertilizers and most farm chemicals.

Tough and highly resistant to puncturing.

Extremely resistant to the transmission of moisture vapour and gases.

High degree of elasticity, it will stretch at least 300%.

If damaged, the sheet can be repaired in a similar manner to mending a bicycle inner tube.

Large 'tailor made' sheets can be fabricated by heat sealing techniques.

Available in a wide range of weights to suit different applications.

Retains its waterproofing properties when fully extended.

Unaffected by the majority of chemicals.

Its durability will not significantly diminish with age.

A comparison of relative permeabilities for four materials is illustrated in Fig. 54.

Although butyl has not been used extensively for lining canals, some test installations have been made and it has given excellent seepage control.[58] Buried butyl linings were installed on the Bureau of Reclamation, Arch Hurley project in the vicinity of Tucumcari, New Mexico, and the W. C. Austin project in Altus, Oklahoma, April 1961.[59] Butyl sheeting, $0·8$ mm ($\frac{1}{32}$ in.) thick, fabricated in full widths of $10·7$ m (35 ft.) and lengths

RELATIVE PERMEABILITY

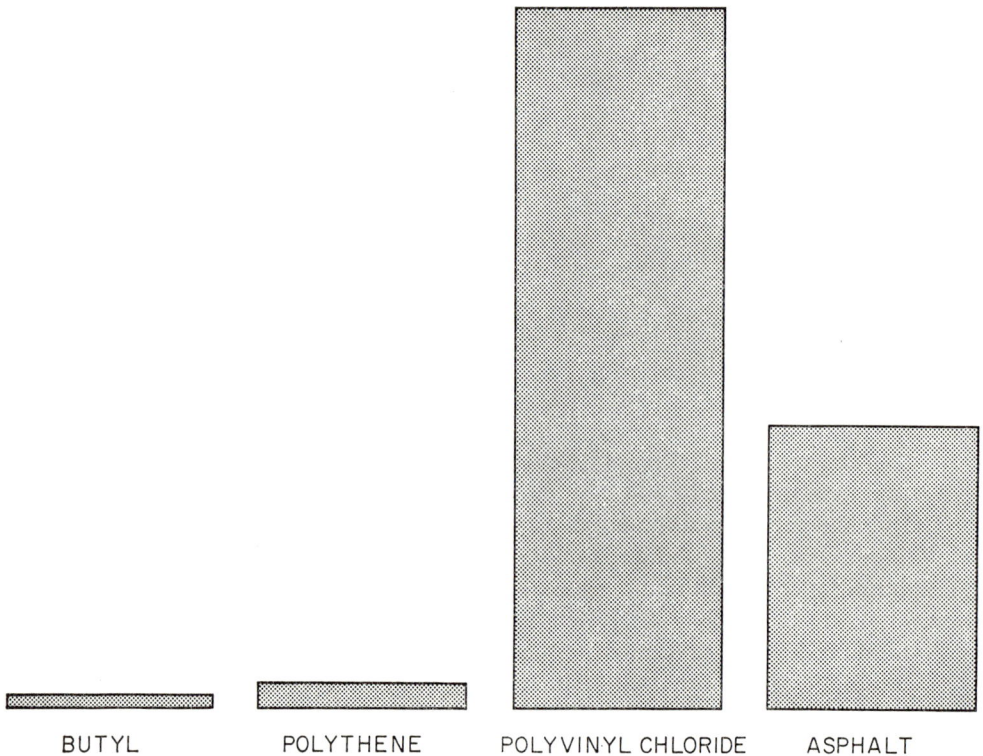

Fig. 54 Comparison of relative permeability of selected materials, actual data as determined by ASTM test method E96–537.

of 62·5 m (205 ft.), was used. The sheeting was folded to make a 3 m (10 ft.) width and wound on a cardboard, 150 mm (6 in) hollow core for handling and shipping. The rolls weighed approximately 636 kg (1400 lb) each. To position, the butyl liner was unrolled on the berm of the canal, with the aid of a crane and a lifting bail. A lifting bail should be used to prevent damage to the ends in handling. The top edge of the liner was anchored in a trench similar to the way reservoir linings are installed. The upstream edge was anchored in a transverse trench, and the downstream edge allowed to overlap the upstream edge of the lining downstream. A canal being lined with prefabricated butyl sheeting is shown in Plate 9.

The use of P.V.C. pipes for drainage of building sites has demonstrated a number of advantages. It is very light, 30 times lighter than clay. A length of 6·10 m (20 ft) of 10 cm (4 in) pipe weighs about 11 kg (24 lb), hence its ease of handling on sites makes special transport unnecessary. Only minimum trench widths are required and cutting on site is simple, and all off-cuts are re-usable. Extension of the use of such pipes in the irrigation/drainage field would appear to be a practical consideration.

Sprinkler irrigation

The 1963 British Act also stimulated interest in supplementary irrigation and in the need to improve efficiencies in water application. This has led to refinements in sprinkler design including portable systems to provide low labour content in irrigation of high value crops, where a small increase in water yield, and in quality, makes a large difference to the grower's income. The flexibility in design makes it possible to cater for a wide range of different needs. Grass leys for instance require regular applications for steady growth to ensure the correct feeding rate for livestock, whereas yield from crops such as wheat, barley, and peas can be increased appreciably by one or two applications at the right times.

Plate 10 illustrates one of the latest developments in sprinkler irrigation.

Metering

Implementation of these practices requires accurate metering and in consequence there have been striking refinements in the manufacture of metering equipment. The range includes open flow; irrigation low pressure saddle; vertical flow; portable-type spray and portable low-pressure meters. Illustrated in Plate 12 is an example of one of the latest free-flowing vertical flow meters.

Nuclear and electrical fields

Development in the nuclear generation fields is such that it can be said that around 1980 the cost of producing energy by using uranium resources will at least match the cost of fossil fired units. In association with advances in desalination techniques these developments also hold great hopes for arid regions in the provision of domestic and agricultural water supplies.

But this will not be achieved quickly nor without intensive research. Although certain fundamental principles will always be associated with the technical aspects of desalinization, the interplay of technical, economic and social factors presents a complex picture, when specific applications are considered. With foreseeable technology economics will determine that nuclear power be restricted to big plants. The report issued by the International Atomic Energy Agency of the joint United States–Mexico Study team on nuclear power and water desalting plants for the Southwest United States and Northwest Mexico considered a plant with 2000 MW and a water output of one milliard gallons (4 540 000 m³) per day.[60] It offered one of the lowest costs of desalinated water yet seen but the range of 9–31 U.S. dollars per thousand gallons was obtained by assuming the *unproven technology of the fast breeder reactor and very low interest rates at 4%.*

The possibilities from a breakthrough in technology for developing small, low cost reactors, of the order of 150 MW electrical with 454 000 m³ (100 million gallons) a day of fresh water would augment handsomely the water supplies of many communities.

In the electrical field the development of the 'solid state' or 'Thyristor' converter systems hold immense promise with respect to long distance, high voltage, direct current transmission of large blocks of power. The Cabora Bassa project is planned to deliver, in the first instance, nearly 2000 MW, over a distance of some 1800 kilometres (1100 miles) from the Zambesi river in Mozambique to Irene, in the Republic of South Africa. This will contribute significantly to the concept of multipurpose development. Thus it will be possible to accelerate the development of remote river basins and to promote regional co-operation through the export/import of large blocks of energy over very long distances, thereby creating early revenue earning opportunities through the sale of electricity which may be utilized to accumulate funds for subsequent agricultural and industrial developments in the remote regions. The possible impact on raising of standards of living in such regions will be far reaching indeed.

Pumps

Both for irrigation and for vertical drainage a wide range of pumps has been developed for river extraction, canal lifting, delivery up an escarpment to irrigated areas, overhead spray irrigation, flood irrigation and for drainage. Plate 11 illustrates a submersible pumping installation.

CONCLUSION

The foregoing examples are by way of illustration and are by no means exhaustive.

It is vitally necessary to make the point that whereas it is appropriate to make a realistic assessment of the *problems* existing in the various parts of the world, *it is also proper to take account of all significant technological advances which are offering aids* to development covering most of the problems to be encountered.

There is a need to dispel any feelings of the new graduate that all is known about the laws of hydraulic engineering in general and of irrigation engineering in particular. On the contrary we are but seeing the drop from which to visualize the ocean. Thus we are but at the beginning of a new wave of endeavour and discovery; an interesting transition stage where the impact of time disciplines is just beginning to be felt. Some classical theories will be advanced—others will be challenged and modified. There is work of interest for all.

In the sense that research is intensive original study we might, in conclusion, refer to the words of Bacon:[61]

> Foolish men condemn studies;
> Simple men admire them;
> Wise men use them.

9 The concept of catchment engineering

On a global basis, the greatest problems in water resources engineering stem from the fact that the scarce resources—water, finance and technical capacity—are not generally sufficient to meet the ever-increasing water requirements for the rapidly expanding populations and rising standards of living. The problems generated through lack of mis-management of water resources are of regional and national concern. As regards nutritive viability of the overpopulated areas, the interests of all mankind are already involved. Thus, engineers have come to realize that successful solutions, in time, will require, firstly, close co-operation between many disciplines besides pure engineering and, secondly, maximum use of all the tools and devices which are becoming available for the rapid ranking selection and implementation of projects.

THE CONCEPT

Integrated planning of river basin development means the assessment of total water resources for a given catchment or basin in relation to adjacent and, in some cases, remote basins, followed by the identification and formulation of those projects having the highest priority, within a given period of projection, for the allocation of scarce resources: finance and implementation capacity. The new trend has grown from the realization that, on the basis of existing criteria for the selection of projects, the single, large prestige project could not in itself fulfil the requirements of optimizing, let alone maximizing, the benefits from available resources. Moreover, apart from the fact that economic criteria have not yet succeeded in taking into effective account the relative length of life of civil engineering structures, experience has shown that any large project is, or may be, vitally affected by developments both upstream and downstream. Upstream considerations cover eventualities such as sudden failures of other structures, the acci-dental maloperation of spillway gates, generation of silt, importation and diversions of water and all factors involving changes in run-off pattern. Downstream considerations relate to changing patterns in water use, depletion of reservoir capacity, allowable flood discharges, the effects of discharging 'hungry' water and re-use potential of water generally.

The steady evolution of the concept of integrated basin planning was assured once it became accepted that the single large storage project cannot be properly planned from the point of view of true multipurpose function, except for a given instant in time, and that the possible interplay of dynamic factors affecting the economics of water supply and demand invariably require a 'wider' look at the basin as a whole. This matter has been covered in more detail in Chapter 3 ('Time disciplines in water resources plan-ning').

The concept is universal and not confined to the developing and/or water deficient regions. Whereas it is the pressing need to achieve and maintain nutritional viability in the overpopulated and generally water deficient regions that provided the impetus for refinement of techniques and procedures, the application is of equal importance to the advanced industrialized and, therefore, built-up regions. In England sizeable sums of money have been earmarked for an intensive study of the causes of flood generation in order to provide guide-lines for meaningful measures, on a basin approach, to mitigate the effects from recurrent and increasing flood intensities and to preserve rural amenities and catchment potential. Apart from studying the obvious causes affecting sensitivity of run-off, such as progressive covering of land surface with impermeable tarmac, concrete and building structures, coupled with diminished forest and hedge cover, attention will be directed to some less obvious but equally insidious elements affecting water regimes. It will, for instance, be deemed worth while to study the long-term effects on runoff and movement of underground water of soil compaction by heavy farm machinery on catchment areas such as the Berkshire Downs, where the earth appears to be growing less porous, its crumb structure impaired by ever heavier use of chemical fertilizers. The effects of reduction of deforestation and the progressive drainage of moorland bogs acting as sponges may have to be countered by 'headwater arrestment' measures. At the other end of the scale is the development of coastal barrages, such as has been studied at Morecambe Bay in Cumberland, designed to husband the residual fresh water of the basin at its lowest point before entering the sea and to defend its quality against encroachment of tidal saline waters.

The following are some of the major factors which are contributing towards the evolution of the wider approach:

(a) *The example set by financing authorities*, such as the World Bank which, during 1966/67, provided some U.S. $877 million of new commitments in loans for projects and studies. In a Paper dealing with methods of the World Bank for the evaluation of projects and the control of their implementation, M. Bernard Chadenet,[62] Director of its Projects Department, stated:

'The Bank is placing more and more emphasis on integrated projects. Whereas the Bank often satisfied itself by financing a dam or the principal works, the works taken in the broad sense comprise not only the large works but also all that is indispensable for their effective utilization: agricultural machinery, road systems, credit and agricultural advice on distribution of fertilizers, seed, price and policy.'

(b) *The benefits flowing from dissemination of research* carried out on a global basis during the post-war era, made possible by vastly improved methods of communication. Thus, it is now possible, on the basis of climatic data, to forecast or precalculate the crop water requirements for any cropping pattern anywhere in the world and thus to study and compare various alternative systems before commitment to a given project. As regards provision for the routing of major catastrophic floods the meteorological-hydrological approach (maximized moisture content over a catchment) provides a sound basis for design and a check on the statistical methods, based on past records of flow. Recent examples of large storage projects where this method was used to determine spill capacities for routing catastrophic floods are to be found in the Mangla Dam in West Pakistan and the Hendrik Verwoerd and P. K. de Roux Dams in the Republic of South Africa.

(c) *Developments in computer programming* during the past decade have made it possible for the first time to undertake, within a reasonable time, the massive volume of calculations required to make a proper analysis of complex irrigation and multipurpose systems in the light of different sources of water supply and taking account of all foreseeable constraints. Having decided upon production objectives, cropping patterns and capabilities of the agricultural community to cultivate and manage the farm, it is possible to build up patterns of water demand for comparisons of water availability. From this it becomes possible to derive preliminary ideas of the structures required to store and to convey the supplies to the fields, and to estimate the correct drainage pattern, vertically and horizontally, to avoid rising water tables and the effects of waterlogging. This involves intensive trial and error procedures, and comparison of numerous alternatives. The computer has the great advantage that, having started at farm level and worked back to the diversion site, there is no need to remain wedded to the first concept because the computer can substitute various assumptions in an attempt to optimize the return on capital invested and the phasing of financing needed.

(d) *Recent development in methods and procedures in economic and financing analyses for evaluation of projects*—a field in which there has been much confusion and disagreement from country to country and as between various authorities. Indeed, this is one of the chief reasons for the loss of 'dialogue' in recent times by engineers owing to the welter of criteria bandied about by economists and other disciplines. The recent trend in engineering economics towards a meaningful approach should do much to redress this situation and to bring the engineer back into the planning and decision-making processes. The criterion that the basic justification of multipurpose and major single-purpose schemes should rest upon *economic* evaluation backed by judgement factors, and that guide-lines for pricing structures should rest upon *financial* analysis, provides a sound basis for meaningful forward planning with respect to water resources projects with special reference to optimum development of basin potential.

(e) *The growing awarness of the need for regional co-operation in the use of all resources*, particularly water, in order to provide the required nutritive viability to ensure maximum efficiency of the nation and at the same time to facilitate the necessary industrial growth rate for expanding standards. This in turn involves energy and power resources. Whereas energy requirements can always be met from large thermal or nuclear units, the essential peak power capacity needed to reconcile demand and supply patterns in an interconnected transmission grid system can still best be supplied from hydro developments which provide the necessary flexibility for instantaneous use. This matter is also referred to in Chapter 5 ('The impact of irrigation on regional economic growth').

(f) *Advances in engineering technology.* The most striking recent advance has been made in rock and earthfill moving plant, which has helped to reduce the cost of mining and placing materials in dams, and it is now possible to construct dams on sites which were previously considered impossible on geological grounds. Recent research has also shown that rockfill dams can now be designed to be overtopped which, in the first instance, will make it possible for rapid and cheap developments in remote upper catchment regions, using mainly materials locally available. This matter is also referred to in Chapter 8.

Two examples are quoted to illustrate the concept:

First example: Indus Basin Project—West Pakistan

The most recent, and possibly the most classic, example of integrated planning of river basin development is the Indus Basin Project of West Pakistan. It is also the best example of an engineering solution to political problems which threatened to disrupt the peace of the Indian sub-continent.

As a result of the Indus Waters Treaty of 1960, India and West Pakistan were given clearly defined rights in respect of the waters of the Indus Basin: three rivers to India and three to Pakistan. A vast system of works was to be constructed over a period of ten years with funds amounting to some £450 million administered by the World Bank on behalf of the six friendly countries which provided the grants. This involved the construction of the Mangla storage dam on the Jhelum, commissioned in 1967, and a system of inter-river link canals some 650 km (400 miles) long, with six new barrages capable of replacing from the western rivers the water to be cut off by India in 1970 from the eastern rivers. The link transfer capacity is equivalent to the entire irrigation requirements of Egypt at present (Fig. 55).

These waters are vital to downstream riparian West Pakistan, whose economy at present depends almost entirely on irrigated agriculture.

Massive as these works are, the effects would in the main be replacement of waters to which Pakistan was entitled in 1947 before partition. The question of development, therefore, was of prime importance.

Thus, the second phase of the approach was a massive study during 1964 and 1965 to investigate the water, and power resources of the Basin. Basically, the approach adopted in the study was to analyse the least cost path of water resources development to maximize agricultural output. However, it was necessary to take account of the power potential having regard to the proven resources of natural gas, some $283 \times 10^{15}\,m^3$ ($370 \times 10^{15}\,yd^3$), and the projected availability of more than 2500 MW of hydro-electric power from the Mangla and Tarbela dams, on the Jhelum and Indus rivers respectively.

Inter-connection of the two power sources proved to be essential to permit the export of energy southwards some 1290 km (800 miles) in periods of high reservoir levels and northwards at the critical periods when reservoirs are emptied for irrigation.

It was found that the extensive groundwater reservoir of some $370 \times 10^9\,m^3$ (300 MAF), twice the mean annual run-off from the basin, could, and should, play a major role in the development of the system. The groundwater reservoir may be used as an overyear storage reservoir, and it was found that considerable economy could be achieved in the design of conveyance structures by relying on the use of groundwater in periods of peak crop water requirements. (This matter is referred to in Chapter 6.)

The tubwell project areas for integrated use with surface supplies are shown in Fig. 56.

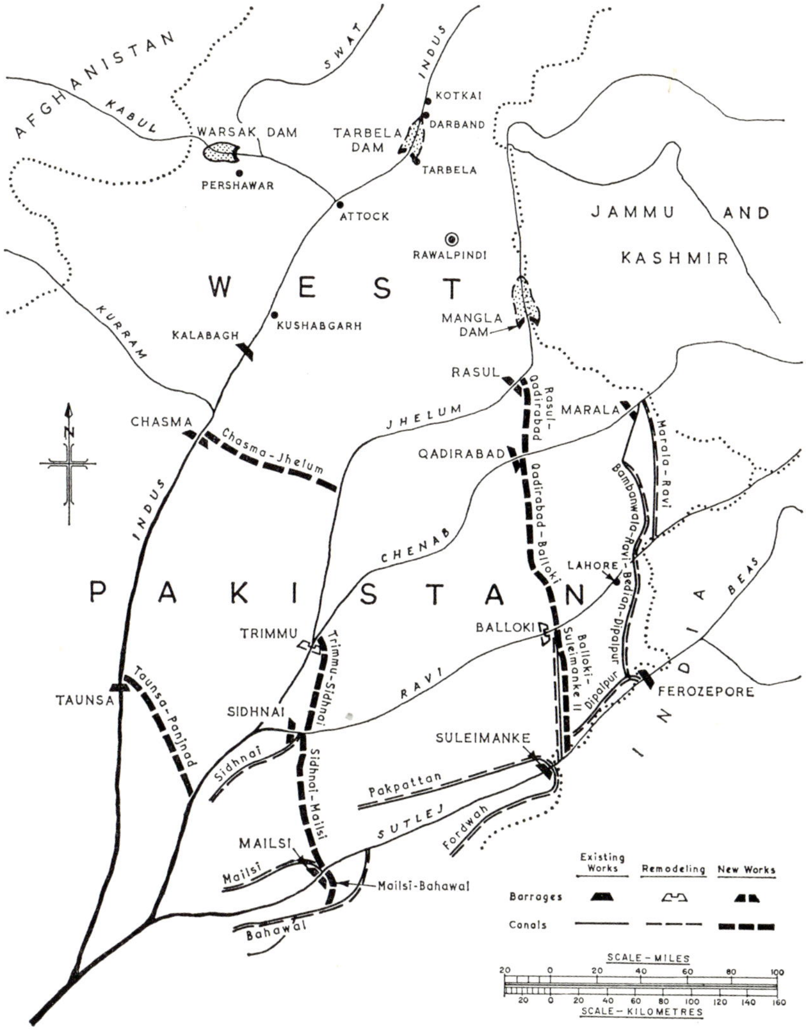

Fig. 55 Indus and Tarbela projects.

One of the major constraints imposed on the freedom of planning of the consultants who undertook the study was the existing irrigation system covering some 162 000 km² (40 million acres) and consuming $98 \cdot 6 \times 10^9$ m³ (80 MAF) of water each year through a conveyance system comprising 61 100 km (38 000 miles) of canals.

It was only possible to undertake the considerable volume of calculation involved in the time available by the use of computers. Indeed, a few years ago, this study could not have been contemplated. Separate computer programmes were written for each stage of the analysis. A prerequisite of such studies is the necessity to obtain a thorough knowledge of the grass roots of the irrigation system, and engineers and agriculturists involved

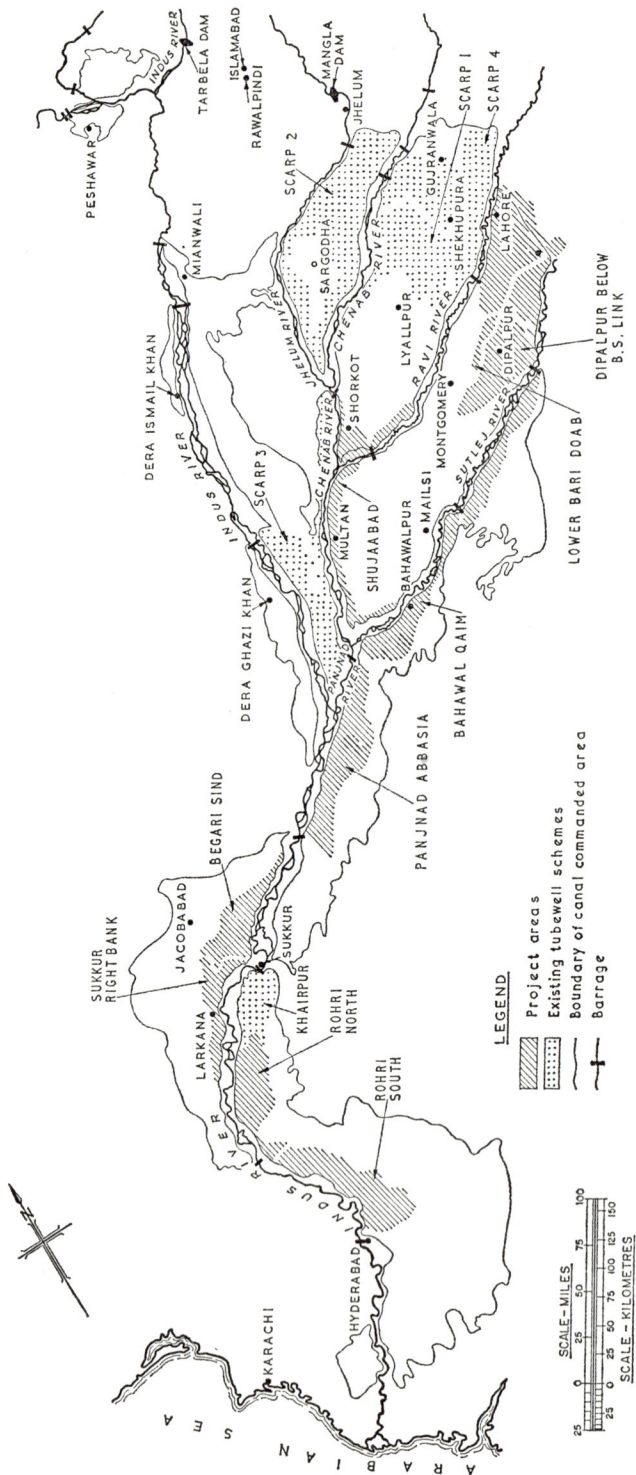

Fig. 56 Integration of ground and surface waters in the Indus Plains.

in the programming must be conversant with the detailed operation of the irrigation systems, both as they exist and as they are likely to develop. This is the only means of ensuring that all constraints are recognized and given due consideration, and various alternatives compared.

The result of the integrated basin approach is that, apart from replacement of lost waters, West Pakistan has planned to achieve a breakthrough from the preservation motivated society to become nutritionally viable around the year 1973–75 on the basis of a national average nutrient level of 2200 calories per head per day. Thereafter, continued water development programmes should result in the economy becoming steadily more profit motivated. (This matter is also referred to in Chapter 5.)

Second example: Spain

The case for integrated basin development in Spain has been clearly stated by Dr. J. Toran.[63] In 1967 Spain had 323 dams completed and 107 under construction. Reservoir capacity amounted to 28 billion cubic metres or 900 cubic metres per inhabitant. Hydroelectric power installed capacity was 8000 megawatts compared with 3500 megawatts of thermal capacity. The area under irrigation was 2·5 million hectares (6·2 million acres).

Spain currently devotes to hydraulic works in public and private sectors some 2·5% of the national income, and more than 20% of national budget.

The tool which Spain needed to activate its extraordinary interest in the expansion of hydraulic works is the recent developments in pumped storage techniques.

The stage is gradually reached in developing countries where transmission networks are adequate to absorb all the energy produced under the load curve, which makes possible the most economic integration of all forms of energy, graded according to availability and efficiency. This has two major economic advantages: export/import relationships between regions become possible whereby surplus power can be supplied for short or long periods of power-hungry neighbouring areas, and it becomes possible to deal effectively with peak loads which, in isolated developments, have to be taken care of by costly, and often inefficient, excess installed capacity. Maximum peak loads arise from weather extremes, industrial activity, and the customs of the population served. A classical modern example of large inter-connection is the energy export/import relationship between countries belonging to the Union for the Co-ordination of the Production and Transport of Electric Power of Western Europe (U.C.P.T.E.) (see Chapter 1).

Spain is a member of U.C.P.T.E. and careful planning to exploit her extraordinarily favourable topographical, geological and hydrological conditions will result in tremendous economic advantages. Consequently, nationwide surveys are in progress to locate and identify reservoir sites on the basis of certain parameters in order 'to apply comprehensively the multipurpose principle to the entire basin'. The objective is not only to develop the resources of the particular basin which, for a single purpose such as power, has a clear-cut solution, *i.e.* the cascade and its pumped storage extensions. It is the comprehensive development leading to complete integration and co-ordination of all the various uses of water with reference to demands from, and supplies to, areas beyond the basin limits.

This concept has led to the conclusion that 'we must overcome the concept of the dam as a singular entity and we have to consider a system of dams as the device leading to the multipurpose and total development of a river basin in connection with other river basin potentials. The aim is to store water not only to regulate its contingent availability due to meteorological conditions, but to provide for a dicontinuous demand.'

Spain is also studying the possibility of large transfers of water between basins in order to adjust the imbalance between zones well endowed with water resources, but with poor agronomic potential, and the areas where the trinome: soil–climate–man displays is at its best and only water is needed for full development.

In any basin the approach for such integrated developments involves the use of small and medium, as well as the very large, dams. In Spain the high altitude sites, the best in Europe, are suitable for the cascade principle and pumped storage. This involves high heads but small storages. Safe and most economic regulation requires large downstream storages in the intermediate catchment zones, *i.e.* just below the mountain cascade systems and above the wide populous valleys, in order to provide safety devices against large discharges, accidental or otherwise, from the upper regions, and to maximize the use for energy and other purposes of spill waters.

The ideal 'solution' is indicated in Fig. 57.

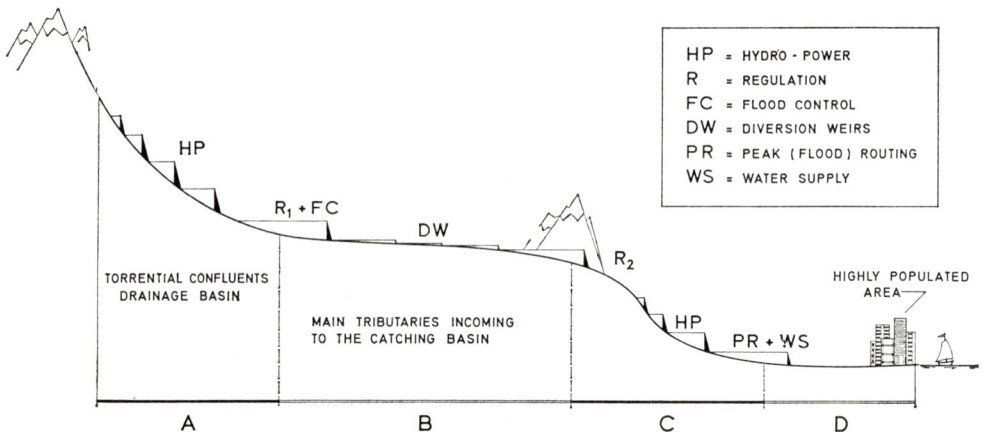

Fig. 57 Diagram of integrated river development.

Fundamentally the multipurpose objective is retained but individual uses are distributed and co-ordinated over a number of sites within the catchment. Generally flood control and regulation criteria are complementary and require large reservoirs without emphasis on head whereas in the case of power consideration of head is vital. The relationship of large 'sponge' reservoirs below the cascade system with arrangements for re-circulation for pumped storage is illustrated diagrammatically in Fig. 57.

It is interesting to note that the first dams for Madrid water supply are more than 100 years old, and are still functioning well, and are to be integrated into the much larger basin systems.

SILTATION OF RESERVOIRS

A matter of great consequence for all countries which have to rely on storage dams for their essential water supplies, is the question of the loss of capacity of the reservoirs due to siltation. The drastic economic effects of depletion of reservoir capacity may best be illustrated by the forecasts for the Tarbela dam in West Pakistan, referred to in Chapter 3. The cost of active storage will be of the order of £28·3 per 1000 m³ (£35 per acre-foot), and it is estimated that active storage will be reduced from $10·6 \times 10^9$ m³ to $1·23 \times 10^9$ m³ (8·6 to 1·0 MAF) in about 50 years. The economic importance of the storage may be gauged from the fact that, for viability studies, the net irrigation benefits, discounted at 8 % to January 1965, were assessed at £250 million and the gross power benefits at £75 million.

The annual silt load of the Indus at Tarbela, from a catchment of 181 000 km² (70 000 square miles), is of the order of 366 million tonnes, most of which is transported during 2½ months of the year.

The major reason for the rapid depletion of the reservoir is the fact that it will retain the sediment from the entire flow of the Indus, averaging $75·2 \times 10^9$ m³ (61 MAF), instead of from the relatively small quantity needed to fill the reservoir. If it were possible to by-pass the reservoir so that only enough water is extracted to fill the reservoir each year, then the life of the storage would be extended by a large factor, up to ten times. However, by-passing a river as large as Indus cannot be done easily and the eventual plans will be based on two different forms of catchment engineering: off-channel storage and attenuation of headwaters. Off-channel capacity to store sediment-free water skimmed from Tarbela reservoir exists in nearby tributary valleys. Headwater control may be achieved by a series of small, relatively cheap dams of the same type as those already being constructed, with interesting results, on the tributaries to the Jhelum above the Mangla reservoir.

In order to protect the live storage capacity of the Mangla reservoir against silt encroachment a programme for construction of small dams on all the tributaries was commenced a few years ago. To date some 300 such dams have been completed with very successful results.

Plate 13 illustrates the effectiveness in silt trapping of one such dam after only two years of operation.

There is some 5·5 m (18 ft) of silt trapped on the upstream side at the face of the little dam which is over 7 m (25 ft) high. It also shows one of the by-products of such catchment protection measures—the lush growth of elephant grass. The fertile growth extends upstream on a 'backwater curve' basis and traps fertile silt even more effectively than the static waters of the reservoir. Within a few years it will be possible to remove the elephant grass and to plough and cultivate the land, as in the case of the Sipiali dam, Plate 14.

Thus, the local cultivators, who were originally opposed to development on the grounds that their lands might be flooded, now enthusiastically demand an increased programme of 'catchment engineering'.

Catchment engineering is thus concerned not only with the large prestige works, but with the whole range of structures which will maximize the benefits from the water yield of the catchment area. In this context, research carried out in association with the British Hydromechanics Research Association's Laboratories[51] between 1963–68 on models of rockfill dams culminated in the discovery of significant parameters which will permit the design of rockfill dams so that they may be overtopped by flood-waters without risk. (This matter is referred to also in Chapter 8.)

The application of these techniques to both temporary and permanent works will assist the engineer to develop rapid methods of dam design and construction. Particular advantages foreseen are that construction may take place on relatively poor foundations using a high proportion of local materials and local, even unskilled, labour.

It will be obvious that the application of these theories to under-developed territories can make a great difference in the rate of development and in the developed regions catchment engineering with these methods may ensure an increasing measure of protection against flood damage to high value properties.

The application of the theories can vary over a wide spectrum of requirements. For example, the construction of side spillways in reservoirs where there are suitable 'saddles' will be effective in reducing the freeboard requirements for flood insurance, thus saving cost of the main dam. The question of reducing spillway capacity in large dams is very important when one considers that in some rockfill dams the cost of the spillway represents upwards of 30 % of the cost of the main dam. Another example is in the construction of temporary cofferdams for major projects.

OUTLOOK

The philosophy of integrated basin development requires comprehensive long-term planning over a wide range of disciplines and therefore automatically involves sound links or effective co-ordination between planning and implementation to ensure that gaps are not likely to develop between cost estimates prepared by planners operating in isolation and the actual ultimate construction costs.

As regards regional planning the accelerating pressures on resources, human and material, with particular reference to water, has led to closer consideration of potential for maximizing benefits from regional collaboration in use of water resources, particularly as international or regional boundaries do not always follow watersheds and often more than one nation or region has riparian rights to a common water supply.

Moreover, there is a distinct trend towards regional co-operation, *i.e.* 'common market' tendencies on the basis of beneficial export/import agreements designed to alleviate scarcities in one part by surplus from another thus ensuring that the benefits from the whole are greater than those from the sum of its parts.

Considerations of such national and regional importance require, in addition to sound engineering and evaluation techniques, adequate administrative and legislative facilities to permit fruitful planning and successful implementation.

The administrative and legislative arrangements vary from country to country. Some major influencing factors are the stage of development attained, the population density and growth rate, geographical distribution of population in relation to climate and multipurpose activities in relation to water resources; and the extent to which water is recognized as a limiting national resource.

In the case of Australia special problems arise from the fact that some 60% of the population is concentrated in four cities. In West Pakistan the overriding influence factor in recent times has been the need to ensure, by water resources development, a breakthrough from a preservation-motivated to a profit-motivated society.

The concept of catchment engineering or integrated basin development deserves special attention from all disciplines and particularly from engineering. It offers the best opportunities for balancing the increasing weight of problems encountered with aids from technological and conceptual advances.

Bibliography

1. GULHATI, N. D. 'World view of irrigation developments.' *Proc. Amer. Soc. civ. Engrs.*, **84**. IR3, Paper 1751 (Sept.) 1958.
2. OLIVIER, H. 'Efficiency of water distribution and use on the land.' *Proc. 5th Congress Int. Comm. Irrig. Drnge.*, Tokyo, R.3, Question 16, 1963.
3. KRUTILLA, J. V. and ECKSTEIN, O. *Multiple Purpose River Development* (Studies in Applied Economic Analysis), Baltimore: Johns Hopkins Press, 1957.
4. MCKEAN, R. N. *Efficiency in Government through Systems Analysis with Emphasis on Water Resource Development* (Publications in Operations Research, N.3), New York: John Wiley & Sons, 1958.
5. NIX, J. F. and PRICKETT, C. H. 'The economic aspects of farm crop irrigation.' *Rep. Frmg. Econom. Camb. Univ.*, No. 55, Univ. of Cambridge School of Agriculture, Cambridge University Press, June 1961.
6. PENMAN, H. L. 'Natural evaporation from open water, bare soil and grass.' *Proc. Roy. Soc.*, **193**, 120–145 (April), 1948.
7. BEAUCHAMP, K. H. 'Potential use of water by irrigation in the humid areas.' *Proc. Amer. Soc. civ. Engrs.*, **84**, IR3, Paper 1750 (Sept), 1958.
8. KERR, R. S. (Chmn.) Select Comm. Nat. Wtr. Res. U.S. Senate Comm. *Print No. 12.*
9. UNITED NATIONS FOOD AND AGRICULTURE ORGANIZATION. *Food & Agriculture Production Yearbook, 1962.*
10. OLIVIER, H. 'Irrigation as a factor in boosting food and fibre production.' *Proc. Nutr. Soc.*, **24** (1), 8–21, 1965.
11. WEST, Q. M. *Food: One Tool in International Economic Development.* Center Agric. Econom. Adjustment, Iowa State Univ., 1962.
12. PAWLEY, W. H. *The Priority to be accorded to Agricultural Development in the National Economic Planning.* Symposium, Roy. Commonwlth. Soc. Agr. Conf. 1965.
13. EDMINSTER, T. W. 'Need for new and creative research in irrigation and drainage.' *Proc. Amer. Soc. civ. Engrs.*, **90**, IR4, 18–23 (Dec), 1964.
14. WOLLMAN, N. *The Value of Water in Alternate Uses.* Albuquerque: University of New Mexico Press, 1962.
15. ANJARWALLA, Z. K. *Technique of Improved Irrigation Efficiencies through Control of Soil Moisture in Relation to Internal Plant Stress Requirements.* M.Sc.(Eng.) Thesis, University of Southampton (Sept.), 1965.
16. SYLVESTER, R. O. and SEABLOOM, R. W. 'Quality and significance of irrigation return flow.' *Proc. Amer. Soc. civ. Engrs.*, **89**, IR3, Paper 3624 (Sept.), 1963.
17. STOECKLER, J. H. 'Shelterbelt influence on great plains field environment and crops.' *U.S. Dept. Agr. Prod. Res. Report*, No. 62, 22 pp., 1962.
18. U.S. FEDERAL POWER COMMISSION. *National Power Survey, 1964.*
19. UNION FOR THE CO-ORDINATION OF THE PRODUCTION AND TRANSPORT OF ELECTRIC POWER OF WESTERN EUROPE. *Annual Report 1964–65.* Secretariat Arnhem, Netherlands (English version by Dr. Ing. L. Wolf of Munich, W. Germany).
20. HUDSON, SIR W. *The Snowy Mountains Scheme.* Official Yearbook, Commonwealth of Australia, No. 42, Govt. Printer, Canberra, Austr., 1956.

21. SOANE, B. C. *Soil Moisture and Plant Growth.* Proc. Int. Irrigation Symposium, Wright Rain Ltd., Salisbury, Rhodesia.

22. OLIVIER, H. *Development of Hydropotential in Relation to Climate.* Ph.D. Thesis, London University, 1953.

23. OLIVIER, H. *Irrigation and Climate.* London: Edward Arnold (Publishers) Ltd., 1961.

24. KEELING, B. F. E. *Evaporation in Egypt and the Sudan.* Survey Department, Paper No. 15, Min. Finance, Cairo, 1909.

25. BRIGGS, L. J. and SHANTZ, H. L. 'Daily transpiration during the normal-growth period and its correlation with the weather.' *I. Agric. Res.*, **5**, No. 14, Washington, U.S.A. (Jan.), 1916.

26. JENSEN, M. E. and HAISE, H. R. 'Estimating evapo-transpiration from solar radiation.' *Proc. Amer. Soc. civ. Engrs.*, **89**, IR4, 1963.

27. BLANEY, H. F. 'Monthly consumptive use requirements for irrigated crops.' *Proc. Amer. Soc. civ. Engrs.*, **85**, IR1, 1959.

28. PETERS, D. B. 'Relative magnitude of evaporation and transpiration.' *Agronomy Journal.*

29. PATRIC, J. H. 'The San Dimas large lysimeters.' *Journal of Soil and Water Conservation*, Vol. 16, No. 1 (Feb.), 1961.

30. SHAW, R. H. *Transpiration and Evapo-Transpiration as related to Meteorological Factors.* Final report, U.S. Weather Bureau, Contract No. Cwb–9295, Dept. of Agronomy, Iowa State College, 1959.

31. BAHRANI, B. and TAYLOR, S. A. 'Influence of soil moisture and evaporation demand on the actual evapo-transpiration from an alfalfa field.' *Agronomy Journal.*

32. GARDNER, W. R. and EHLIG. 'Relationship between transpiration and the internal water relations of plants.' *Agronomy Journal*, Vol. 56, No. 2, 1964.

33. SIR ALEXANDER GIBB AND PARTNERS (LONDON). *Iraq–Kuwait Water Scheme.* Nov., 1965.

34. RUTTAN, V. W. *The Economic Demand for Irrigated Acreage.* (Resources for the Future Inc.), Baltimore: Johns Hopkins Press, 1965.

35. *Introduction of Supplemental Irrigation Water.* Tech. Bull. No. 76, Colorado State University, Fort Collins, Colorado, U.S.A., June 1965.

36. Gianinni Foundation Research Report No. 257, Sept. 1962.

37. U.S.D.A. *Agricultural Statistics. 1962.* Govt. Printing Office, Washington, 1963.

38. CHRISTENSEN, R. P. and STEVENS, D. R. *Food: One Tool in International Economic Development.* Iowa State University Center for Agricultural and Economic Adjustment, Iowa State University Press, 1962.

39. SIR ALEXANDER GIBB AND PARTNERS, I.L.A.C.O., N.V. AND U. Hunting Technical Services Ltd. 1966. *Programme for the Development of Irrigation and Agriculture in West Pakistan.*

40. THOMAS, R. O. *J. Irr. Drge. Div., Proc. Amer. Soc. civ. Engrs.*, **84**, Paper 1754.

41. GEBHARDT, D. S. *Urban Water Requirements in Johannesburg.*

42. *Benefits and Costs.* U.S. Dept. of the Interior. Bureau of Reclamation Manual, Vol. XIII, March 1952.

43. *Animas–La Plata Project, Colorado and New Mexico.* Letter from the Secretary of the Interior to the Speaker of the House of Representatives, Washington, D.C., May 1966.

44. WALKER, A. L., *et al. The Economic Significance of Columbia Basin Project Development.* U.S. Bureau of Reclamation, Dept. of the Interior, 1966.

45. ALLANSON, B. R., HENZEN, M. R. and COETZEE, O. J. *Environmental Factors in Relation to Water Use and Protection Against Pollution.* C.S.I.T. Effluents Conference, Pretoria, 1964.

46. BLISS, J. H. 'Water quality changes in Elephant Butte reservoir.' *J. Irr. Drge. Div., Proc. Amer. Soc. civ. Engrs.*, **89**, Paper 3637.

47. BASOV, G. F. 1941. 'The influence of shelterbelts of the Kamennaya Steppe on the regulation of surface run-off.' *Lesno Khozyaystvo*, **2**: 8–18 (Russ.) (*Forestry Abs.*, 6: 140, 1945).

48. MOLCHANOV, A. L. 1956. 'Air-temperature, air-humidity and soil-temperature regimes in the fields between shelterbelts.' *Trudy Kazakh. gidrometeorol. Inst.*, **7**: 57–65 (Russ.) (*Soils & Fert. Abs.*, **20** (2): 89, No. 485, 1957).

49. SOKOLOVA, N. S. 1937. 'Influence of shelterbelts on the yield of agricultural plants, grains, products of truck farming and oil-bearing plants.' *Polezaschtchitnye Polossy* (Shelterbelts: Report of All-Union Sci. Research Inst. for Improvement of Farmland by Forestation), **8**: 120–158 (Russ.). Transl. by C. P. de Blumenthal, *U.S. Dept. Agr. Library Transl.*, No. 7317, 68 pp., 1937.

50. STEUBING, L. 1952. 'Dew and the influence of windbreaks on it.' *Biol. Zbl.*, **71** (5/6): 282–313 (Ger.) (*Forestry Abs.*, **14**: 53, No. 349, 1953).

51. IIZUKA, H. 1950. 'Experiment on model windbreak (2nd report): effect on evaporation.' *Meguro For. Expt. Sta. (Tokyo) Bul.*, **45**: 17–26 (Jap.) (*Forestry Abs.*, 15: 48, No. 286, 1954).

52. STAPLE, W. J. and LEHANE, J. J. 1955. 'The influence of field shelterbelts on wind velocity, evaporation, soil moisture and crop yield.' *Canad. Jour. Agr. Sci.* 35: 440–453, Illus.

53. KARUZIN, B. V. 1936. 'A study of the narrow shelterbelts in the Timashev District of the Kuibyshev region.' *Polezashtchitnye Polossy* (Shelterbelts: Experiments of All-Union Res. Inst. for Improvement of Farmland by Forestation), 6: 117–169 (Russ.), Transl. by C. S. Beliavsky, *U.S. Dept. Agr. Library, Transl.*, No. 7311, 33 pp., 1937. Also KARUZIN, B. V. 1947. 'Measures to obviate poor grain quality in the South East.' *Sovet. Agron,* 2: 34–38 (Russ.) (*Field Crop Abs.*, 1: 142, No. 743, 1948).

54. BATES, C. G. 1944. 'The windbreak as a farm asset.' *U.S. Dept. Agr. Farmer's Bul.*, 1405 (rev), 22 pp., Illus.

55. MATZUI, Z. and YOKOYAMA, C. 1955. 'On the efficiency of the windbreak hedge in protecting the rice crop of Horumui District.' *Hokkaido Forest Expt. Sta. Spec.*, Report No. 3, pp. 168–177 (Jap.) (*Forestry Abs.*, **16**: 515, No. 4103, 1955).

56. SOEGAARD, B. 1954. 'Outline of shelterbelts and shelterbelt tests in Denmark.' *IUFRO 11th Congr. Proc.*, Sec. 11, No. 3, p. 269. Illus. Rome, 1953. (Eng.) (*Forestry Abs.*, **15**: 314, No. 2523, 1954).

57. OLIVIER, H. 'Through and overflow rockfill dams—new design techniques.' *Proc. Instn. civ. Engrs.*, Vol. 36, Paper No. 7012, March 1967.

58. *New Materials and Methods for Water Resource Management.* A Report to the Committee on Public Works, U.S.D.A., United States Senate, 87th Congress. Comm. Print No. 6, 1962.

59. *Linings for Irrigation Canals.* Bureau of Reclamation, U.S. Department of the Interior, 1963.

60. QUARRINGTON J. A. *Nuclear Desalination—A Challenge for an Economic Small Reactor.* XIVth Nuclear Congress, Rome, 1969.

61. BACON, F. (Lord Verulam 1561–1626) *Essays.* London: J. M. Dent & Sons Ltd.

62. CHADENET, M. B. 'Méthodes de la B.I.R.D. pour l'évaluation des projets et la contrôle de réalisations.' *Sciences et Techniques*, Nos. 8 and 9, 1967/68.

63. TORAN, J. *Revista de Obras Publicas 1967.* Ninth Int. Congr. on Large Dams, Istanbul.

SUPPLEMENTARY REFERENCES RELATING TO CHAPTER FOUR

(a) *Proposed Practices for Economic Analysis of River Basin Projects*. Report to the Inter-Agency Committee on Water Resources by the Sub-Committee on Evaluation Standards, May 1950; revised May 1958. Government Printer, Washington, D.C.

(b) KELSO, M. M. *Evaluation or Secondary Benefits of Water Use Projects*. Report No. 1 *Research Needs and Problems*. Proceedings of the Committee on the Economics of Water Resources Development of the Western Agricultural Economics Research Council at Berkeley, California, 1958.

(c) MCKEAN, R. N. *Efficiency in Government through Systems Analysis—with Emphasis on Water Resource Development*. Rand Corporation Research Study, 1958.

(d) ECKSTEIN, O. *Water Resource Development—the Economics of Project Evaluation*. Harvard Economic Studies, Vol. CIV, Cambridge, Mass., 1961.

(e) KNEESE, A. V. *Water Pollution, Economic Aspects and Research Needs*. (Resources for the Future, Inc.), Washington, D.C., 1962.

(f) MAASS, A., HUFSCHMIDT, M. M., DORFMAN, R., THOMAS, H. A., MARGLIN, S. A. and FAIR, G. M. *Design of Water Resource Systems*. Harvard University, Cambridge, Mass., 1962.

(g) KEMBALL, N. D. and CASTLE, E. N. *Secondary Benefits and Irrigation Project Planning*.

(h) WANTHRUP, S. V. C. *Benefit Cost Analysis and Public Resource Development*. (Economic Evaluation Concepts), Iowa State University, 1965.

(i) HARRAL, C. G. *Preparation and Appraisal of Transport Projects*. The Brooking Institution, Washington, D.C., 1965.

(j) FREUND, R. A. and TOLLEY, G. S. *Operational Procedures for Evaluating Flood Protection Benefits*. Iowa State University, 1965.

(k) LIND, R. C. 'Flood control alternatives and the economics of flood protection'. American Geophysical Union, *Water Resources Research*, Vol. 3, No. 2; 1967.

(l) BOULDING, K. E. *The Economist and the Engineer: Economic Dynamics of Water Resource Development*. Iowa State University, 1965.

(m) FOX, I. K. and HERFINDAHL, O. C. *Attainment of Efficiency in Satisfying the Demands for Water Resources*. (Resources for the Future, Inc.), Washington, D.C., 1964.

(n) CLAWSON, M. *Methods of Measuring Demand for and Value of Outdoor Recreation*. (Resources for the Future, Inc.), Washington, D.C., 1959.

(o) HIRSCHLEIFER, J., DE HAVEN, J. C. and MILLIMAN, J. W. *Water Supply Economics, Technology and Policy*. University of Chicago, 1960.

(p) KRUTILLA, J. V. and ECKSTEIN, O. *Multipurpose River Development: Studies in Applied Economic Analysis*. (Resources for the Future, Inc.), Washington, D.C., 1958.

(q) HAVENMAN, R. H. *Water Resource Investment and the Public Interest*. Vanderbilt University, 1965.

Appendix

The following tables correspond to those containing metric values in the text. These non-metric equivalents are tabulated to assist readers who are not, as yet, familiar with the use made of metric units in this book.

Table 1(a)

COMPARISON OF CROPLAND AND PASTURE OUTPUTS FOR IRRIGATED AND NON-IRRIGATED LAND

| | Value of production per acre in 1947–49 dollars 1954 statistics | | | |
| | Cropland output per acre | | Pasture output per acre | |
Location	Irrigated	Non-irrigated	Irrigated	Non-irrigated
Eastern U.S.A.	336	64	18·83	3·69
Western U.S.A.	145	32	24·18	1·21
U.S.A.	156	50	23·96	1·79

Table 2(a)

DIET LEVELS IN RELATION TO EXISTING AND PROJECTED IRRIGATED AREAS FOR SELECTED COUNTRIES

Country and projection period		Population × 10⁶	National mean daily diet levels kCal/head	Total area irrigated acres × 10⁶
Egypt	1961	26·6	2530	6·1
	2000	56·0	1800	8·4
Sudan	1960	12·1	2500	2·0
	2000	25·0	2500	4·5
India	1960	442·0	2040	58*
	2000	670·0	2300	109*
West Pakistan	1960	43·0	1970	17†
	2000	65·0	2300	32†
U.S.A.	1960	184‡	3100‡	41‡
	2000	311‡	3100‡	280‡
Mainland China	1960	640	1900	183
	2000	1380	2000	453

Notes: * Includes a considerable proportion of double-cropped acres
   † Includes a considerable proportion of double-cropped acres and relates only to areas irrigated by perennial canals
   ‡ High projection

Table 3(a)
TYPICAL COSTS OF STORAGE

Country/region	Project	Storage MAF	Cost £ million sterling	Cost per acre-ft £
Australia, N.S.W.	Snowy Mountains	1·800*	48·0	27·0
Australia, N.S.W.	Burrinjuck	0·185†	4·0	21·5
Australia, N.S.W.	Burrendong	1·361	14·6	10·7
Australia, Queens	Coolmunda	0·061	1·9	31·5
Australia, S.A.	Chowilla	4·750	11·2	2·4
Pakistan, Jhelum	Mangla	4·750	165·0‡	35·0
Central Africa, Zambesi	Kariba	36·000**	31·3‡‡	0·87

Notes: * Annual stored/re-regulated waters available and costs attributable to irrigation
       † Raised dam
       ‡ Costs excluding provisions for power
       ** Operating capacity excluding provision for flood storage
       ‡‡ Costs excluding provision for power

Table 4(a)
ANNUAL IRRI ATION WATER REQUIREMENTS PER ACRE, BY WATER RESOURCE REGIONS, U.S.A., 1954

Water resource region	On farm			Stored/Diverted		Estimated recovery of losses %	Total irrigation require- ment Acre-in
	Net required by plant Acre-in	Efficiency of application %	Total required Acre-in	Storage and delivery efficiency %	Total required Acre-in		
Eastern							
New England	6	60	10	60	17	20	15
Delaware and Hudson	8	60	13	60	22	20	19
Chesapeake Bay	10	60	17	60	28	20	24
South-east	13	60	22	60	37	20	34
Eastern Great Lakes	8	60	13	60	22	20	19
Western Great Lakes	10	60	17	60	28	20	24
Ohio	10	60	17	60	28	20	24
Cumberland	8	60	13	60	22	20	19
Tennessee	9	60	15	60	25	20	22
Upper Mississippi	12	60	20	60	33	20	29
Lower Mississippi	13	60	22	65	34	20	30
Lower Missouri	12	60	20	65	31	20	27
Lower Arkansas, White, Red	12	60	20	65	31	20	27
Western							
Upper Missouri	13	45	29	40	73	55	40
Upper Arkansas, White Red	16	50	32	55	58	55	35
Western Gulf	13	50	26	60	43	55	27
Upper Rio Grande-Pecos	23	40	57	55	104	55	60
Colorado	31	45	69	55	126	55	74
Great Basin	21	45	47	55	85	55	50
Pacific North-west	17	40	42	60	70	60	38
Central Pacific	26	50	52	50	104	55	61
South Pacific	28	50	56	50	112	55	66

Table 5(a)

COMPARISON OF NET RETURNS AND MARGINAL VALUE OF PRODUCTS WITH INCREASES IN IRRIGATION WATER SUPPLIED TO A GIVEN CROPPING PATTERN

Water		Net returns		Marginal value of products per added acre-foot		Crops	
Acre-feet		Dollars		Dollars			
Added	Total	Added	Total	Water @ $3·00	Water @ $0·00	Name	Acres
895	895	48 862	49 862	55·71	58·71	Cotton	200
345	1 240	13 981	63 843	40·52	43·52	Cotton B. beans Total	200 160 360
376	1 616	5 025	68 868	13·36	16·36	Cotton B. beans Alfalfa Total	200 160 61 421
1146	2 762	15 285	84 152	13·34	16·34	Same Alfalfa Total	421 181 602

Table 6(a)

PRINCIPAL IRRIGATION AREAS OF THE WORLD[9]

(Land receiving water by irrigation schemes excluding inundation areas)

Order	Country	Millions of acres	Order	Country	Millions of acres
1	China	183	16	Chile	3·5
2	India	58	17	Peru	3·0
3	U.S.A.	37·7	18	Korea	3·0
4	Pakistan	27·4	19	Australia	2·1
5	Russia	17·8	20	Philippines	2·0
6	Indonesia	14·8	21	Sudan	2·0
7	Iran	11·6	22	Madagascar	1·8
8	Mexico	10·4	23	Viet Nam	1·5
9	Iraq	9·1	24	South Africa	1·5
10	Japan	8·4	25	Syria	1·4
11	Egypt	6·1	26	Morocco	1·3
12	Turkey	4·9	27	Burma	1·3
13	Spain	4·6	28	Colombia	1·2
14	Thailand	4·1	29	Formosa	1·2
15	Argentina	3·7	30	Greece	1·1

G

Table 7(a)

DIET LEVELS IN RELATION TO EXISTING AND PROJECTED IRRIGATED AREAS

Country and projection period		Population ($\times 10^6$)	National mean daily diet levels (kCal/head)	Total area irrigated (acres $\times 10^6$)
Egypt	1961	26·6	2530	6·1
	2000	56·0	1800	8·4
Sudan	1960	12·1	2500	2·0
	2000	25·0	2500	4·5
India	1960	442·0	2040	58*
	2000	670·0	2300	109*
West Pakistan	1960	43·0	1970	17†
	2000	65·0	2300	32†
U.S.A.	1960	184	3100	41
	2000	311‡	3100‡	280‡
Mainland China	1960	640	1900	183
	2000	1380	2000	453

Notes: * Includes a considerable proportion of double-cropped acres
     † Includes a considerable proportion of double-cropped acres and relates only to areas irrigated by perennial canals
    ‡ High projection

Table 8(a)
SUMMARY YIELD PROJECTION FOR DIFFERENT INPUT LEVELS
(Maunds/acre)

Crop	Cotton			Fodder			Wheat		
Input level	L	M	H	L	M	H	L	M	H
Present yield	8·0	8·0	8·0	265	265	265	12·5	12·5	12·5
1974 yield	10·0	10·0	10·0	340	340	340	14·9	14·9	14·9
2024 yield	16·0	30·8	34·2	510	790	860	21·0	40·8	43·7
Average yearly rate of increase in yield 1974/ 2024	1·2%	4·1%	4·8%	1%	2·6%	3·1%	0·8%	3·5%	3·9%

Notes: L: Low input level
M: Medium input level
H: High input level

Table 9(a)
ESTIMATED FUTURE WATER USE IN THE U.S.A.[40]

Year	Water use MAF				Total	Percent of total supply
	Irrigation	Domestic and municipal	Electric power	General Industry		
1955	90·7	12·1	67·0	45·5	215·3	16·3
1965	94·5	16·0	83·0	55·0	248·5	18·8
1975	110·0	20·0	10·0	68·0	298·0	22·6
1995	311·0	28·0	145·0	92·0	576·0	43·7

Table 10(a)

DERIVATION OF INDIRECT IRRIGATION BENEFITS FROM SUMMARY OF FARM BUDGETS FOR ENTIRE PROJECT
(annual values under full development)

Item	With irrigation	Without irrigation	Difference	Factor	Indirect benefit
1. Type of farm	Irrigated farms	Dry farms & grazing			
2. Number of farms	100	10	(90)		
3. Acres per farm	160	1 600			
4. Irrigable acres	16 000	16 000			
5.	Sales to local wholesale and retail business				
6. Fruit and vegetables	$50 000	—	$50 000	5	$2 500
7. Hay and forage	300 000	$10 000	290 000	5	14 500
8. Sub-total, Benefit A	$350 000	$10 000	$340 000		$17 000
9.	Sales for local and non-local processing, marketing, etc.				
10. Grain	$25 000	$50 000	– $25 000	48	– $12 000
11. Fruit and vegetables	25 000	—	25 000	24	6 000
12. Sugar beets	250 000	—	250 000	26	65 000
13. Seed crops	10 000	—	10 000	10	1 000
14. Dry beans	10 000	—	10 000	23	2 300
15. Soybeans	5 000	—	5 000	30	1 500
16. Livestock (meat)	550 000	25 000	525 000	11	57 750
17. Wool	75 000	5 000	70 000	78	54 000
18. Dairy products	25 000	5 000	20 000	7	1 400
19. Poultry products	25 000	5 000	20 000	6	1 200
20. Sub-total, Benefit B	$1 000 000	$90 000	$910 000		$178 750
21.	Purchases for family living and production expenses				
22. Direct farm benefit			$112 500		
23. Less increased perquisites			– 70 000		
24. Increased purchases for family living			42 500		
25. Increased farm production expenses			840 000		
26. Sub-total, Benefit C			$882 500	18	$158 850
27. Total indirect Benefits, A, B and C					$354 600

Note: Prior to 1967 £1 = $2·81
After 1967 £1 = $2·40

Table 11(a)
WEST PAKISTAN: CENSUS LAND USE AND CROPPING INTENSITIES[39]

		Farm size			
		Under 5 acres	5–24·9 acres	25 acres and over	Total
Culturable area		428	2532	1671	4631
Cultivated area	× 1000	397	2337	1467	4201
Net sown area	acres	343	2042	1246	3631
Total cropped area		418	2460	1458	4337
Land use intensity		80	81	75	78
Cropping intensity		122	120	117	119

Table 12(a)
FARM ACCOUNTS IN THE PUNJAB (WEST PAKISTAN)[39]

	Owner Farmer 1954–1955 Pak. Rupees	Tenant Cultivator, 1954–1955 Pak. Rupees
Number of farms	8	9
Area in acres	144·00	121·00
Cultivated area in acres	125·50	118·00
Cropped area in acres	149·89	134·16
Permanent labour family members	15	16
Hired	5	2
Number of pairs of bullocks	14·5	11
Gross income per cultivated acre in Rs.	242·61	180·03
Actual expenditure per cultivated acre in Rs.:		
Manual labour		
Permanent	16·87	2·13
Casual	11·51	5·67
Bullock labour	31·58	24·68
Seed	8·37	8·72
Implements	3·22	2·78
Artisans	3·00	1·70
Manure	1·19	—
Rent	—	62·52
Land revenue taxes and water charges	9·10	3·11
	84·84	111·31
Net income per cultivated acre in Rs.	157·77	68·72

Note: 1954–55 £1 = Rs. 13·33
        1967 £1 = Rs. 11·5

Table 13(a)

DISCOUNTED VALUE OF NET AGRICULTURAL BENEFITS AND UNIT VALUES OF STORED WATERS[39]

Discount rate %	Input levels	Benefits Rs. × 10⁹	Value of stored water Rs./acre-foot
4	Low	4·4	20·5
	Moderate	5·8	26·9
	High	7·7	36·0
6	Low	2·6	12·0
	Moderate	3·4	15·9
	High	4·5	21·0
8	Low	1·6	7·5
	Moderate	2·1	9·7
	High	2·8	12·9

Table 14(a)

WEST PAKISTAN: NET OUTPUT OF LIVESTOCK PER CROPPED ACRE

Gross output per cropped acre		Rs. 65·00
Costs per cropped acre:		
Feed	Rs. 22·00	
Concentrates and salt	5·00	
Hired labour	2·00	
Veterinary	2·50	
Miscellaneous	1·00	
Interest and depreciation	10·00	
Total:		Rs. 42·50
Net livestock output per cropped acre		Rs. 22·50

Table 15(a)
WEST PAKISTAN: FOOD REQUIREMENTS FOR A 2500 CALORIE/DAY DIET

Description		1965	2005
		in millions	
Population:		51·90	125
Adult male units:		41·50	100

	Standard consumption per adult per year in tons	Food requirements in thousand tons	
Cereals	0·1657	6872	16 560
Pulses	0·0311	1290	3 110
Sugar and gur	0·0207	859	2 070
Vegetables	0·0621	2577	6 210
Fruits	0·0311	1290	3 110
Fat and oils	0·0155	645	1 550
Milk and milk products	0·0828	3436	8 280
Fish, meat and eggs	0·0259	1075	2 590

Table 16(a)
EFFECT OF CALORIE
INTAKE ON WORK OUTPUT

Calories available for work	Output (tons)
1200	6·7
1600	9·4
2000	10·8

Table 17(a)
PROJECTED IRRIGATION WATER REQUIREMENTS IN THE U.S.A.: 1954 AND 2000

Description	1954	2000
Irrigated acres	29 552 000	55 512 000
Storage and diversion (acre-feet)	194 564 000	198 606 000
Irrigation (acre-feet)	114 697 000	137 241 000
Application efficiencies		
Eastern Region (%)	60	70
Western Region (%)	45	60

Author Index

Subject Index